Mary Olmstead Stanton

Physiognomy

A practical and scientific treatise

Mary Olmstead Stanton

Physiognomy
A practical and scientific treatise

ISBN/EAN: 9783337257927

Printed in Europe, USA, Canada, Australia, Japan

Cover: Foto ©berggeist007 / pixelio.de

More available books at **www.hansebooks.com**

PHYSIOGNOMY.

A PRACTICAL AND SCIENTIFIC TREATISE.

BEING A MANUAL OF INSTRUCTION IN THE KNOWLEDGE OF

THE HUMAN PHYSIOGNOMY AND ORGANISM,

CONSIDERED CHEMICALLY, ARCHITECTURALLY, AND MATHEMATICALLY;

Embracing the Discoveries of Located Traits, with their Relative Organs and Signs of Character, together with the Three Grand Natural Divisions of the Human Face.

BY MARY OLMSTEAD STANTON.

SECOND EDITION.

PRINTED FOR THE AUTHOR.

SAN FRANCISCO:

SAN FRANCISCO NEWS COMPANY, 413 AND 415 WASHINGTON STREET.

1881.

Argonaut print.

D. Hicks & Co., Bookbinders.

TO ERNST HAECKEL,

PROFESSOR IN THE UNIVERSITY OF JENA,

THIS VOLUME IS DEDICATED WITH THE ADMIRATION AND VENE-
RATION OF

THE AUTHOR.

PREFACE.

To the Reader:

In sending forth this little work to the public, I am impelled thereto by my desire to benefit the masses of mankind in a manner which I believe they very much need. Man's knowledge of himself seems not to have kept pace with the knowledge of his surroundings. It is time, therefore, that there should be an accordance of intelligence between the two, in order that, through Man's comprehension of his powers and possibilities, he may by scientific methods assist in improving his own life, and in perpetuating a race which shall be an improvement on the present one. This can come only through a knowledge of Anatomy, Physiology, Physiognomy, and Hygienic Law, practically applied. I have endeavored to put this science in as plain and simple language as possible, so that the non-scientific reader should not be confused by terms whose meanings might be ambiguous.

The method of classification used in this system of science is in accord with that observed by all naturalists in their classifications of the lower animals, and is based on the *forms* of the human organism which are produced by the intermingling of the Vegetative, Thoracic, Muscular, Osseous, and Brain and Nerve systems. These are treated in the order of evolution—from the first evolved to the latest acquired, the true and perfected cerebral system.

"Scientific Physiognomy" gives the most comprehensive theory of the Mind of any work hitherto presented to the world. It takes the position that mind inheres in the *entire* organism, and that the brain is only *one* source of the mind. This view is supported by Herbert Spencer, George H. Lewes, Ernst Haeckel, Dr. Maudsley, Dr. J. Lauder Lindsay, and all of the most advanced students of Mind and Physiology. In this system this theory has been elaborated and carried to its ultimate by proofs which I believe to be incontrovertible. I would like to give many illustrations of animal life, but the limits of this work forbid. If this system of Physiognomy find a welcome in the minds of the people, I shall be encouraged to give to the world a volume of more extended research.

My hearty thanks are due to the engraver, Mr. Durbin Van Vleck, for his artistic and intelligent treatment of the wood-cuts in this volume.

The faculties of Secretiveness, Force, and Resistance were inadvertently enumerated on page 26 among those of the Vegetative division. Their proper place is in the Architectural division.

Earnest and religious regard for the welfare of humanity has impelled the writing of these ideas. With the hope that they may lead to a correct knowledge of Man, and that this knowledge may conduce to his welfare, physically, morally, intellectually, and religiously,

I am, sincerely, your friend,

THE AUTHOR.

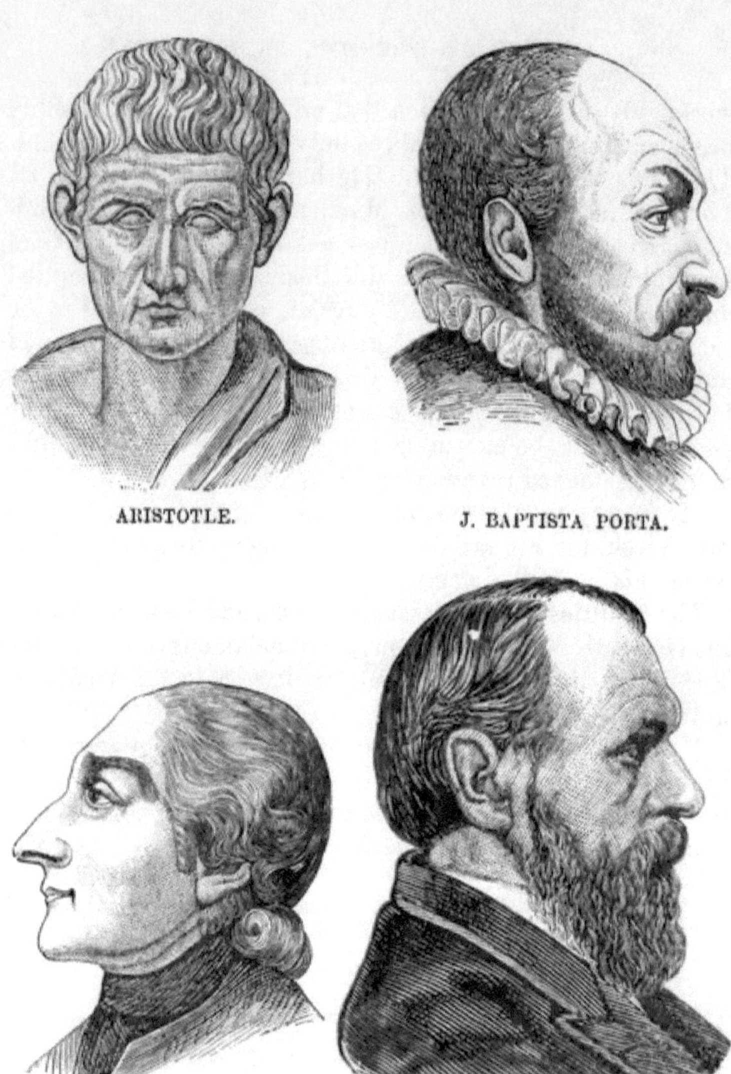

ARISTOTLE.

J. BAPTISTA PORTA.

JOHN CASPAR LAVATER.

DR. J. SIMMS.

FOUR OF THE MOST NOTED PHYSIOGNOMISTS OF ANCIENT
AND MODERN TIMES.

PHYSIOGNOMY.

INTRODUCTION.

If the most learned man of the twelfth century were to
return to earth and become cognizant of our advance in the
sciences and industrial arts, he would doubtless believe, at
first, that he was in the midst of works of magic more wonderful and powerful by far than the mysterious and occult
operations of the Magi of his own age. He would note the
use of natural forces turned to the economies of life by ingenious and complicated machinery; he would be shown the
wonders of steam navigation, of the art of printing, of electricity in its numerous developments and uses, of the tele
graph and the telephone, together with the telescopic and
microscopic discoveries which astonish even this progressed
age; the knowledge of the laws of sound, motion, light, and
color, which this epoch has evolved, would unfold to his
senses a world of realities as new to his mind as if he, in
verity, were transported to quite another planet than the one
which had been his former habitation. After taking note of
all our increased knowledge of Science in its various departments, and after examining our museums and institutions of
learning, if he were to ask, "What do you now know of
Man?—of his powers and properties?" what reply could we
make? We might answer that we understand the circulation

2

of the blood, a little about the nervous system, somewhat of the process of digestion; that we know the number of the bones and have named them, and also the action of the muscles; that we are in a state of uncertainty as to the function of the brain; that we know very little of the prevention of disease, much less about its cure, and nothing at all as to the meaning of his physiognomy. What think you would be his opinion of our progress in useful knowledge? Surely, he would conclude that we had vexed our minds with many things that could be dispensed with, and had neglected the most useful of them all. The knowledge of Man, and how to improve his capacities, how to protect his bodily powers, how to prevent and remedy the diseases which assail him, is surely of more importance than many of the studies upon which valuable time has been spent without advancing the knowledge of Man one step. All through the ages of which we have any recorded history we find inklings of a dawning perception of Physiognomy.

The writings of Moses show him to have been a profound student of human nature and possessed of a power to read and understand countenances and features. His knowledge of sanitary law, in regard to food and diet, and the protection of the body, and the success attending the application of these laws, place him even beyond the Sanitarians of to-day. Among the earliest Greek writers, Aristotle, Plato, and Galen may be named as having written and taught Physiognomy. Hippocrates also formulated a system based upon the several colors of the human complexion. This classification has passed down to the present day, and has been accepted by naturalists in its application to man, while at the same time, with singular inconsistency, the lower animal kingdom has been classified on the basis of form; and correctly so, as color is an effect, not a cause; it is dependent on climate, food, habit, and other accidental surroundings. Even Phrenologists, who ought to know better (since their researches extend widely among the animal kingdom), have retained the classification which Hippocrates set up. The differences observable in the human family he denominated Temperaments—a word which has no intelligent application even to the false basis upon which the old Greek physician founded his system, long before the circulation of blood was discovered by Harvey, and before the functions of the liver, heart, and brain were at all understood.

Each age has added its contributions to our knowledge of Physiognomy, and if these contributions have not given us heretofore a correct system, at once practical and scientific, they have maintained an interest and a belief in this science. This interest and belief have served as a beacon-light, which has flashed far down the ages made brilliant by the works of the most renowned philosophers and literates. Among the Grecians, Aristotle wrote extensively on this subject. Pliny, Cicero, and others of ancient Rome, found this science worthy of their consideration; while, later in the advancing centuries, we find Petrus d'Abbano lecturing on Physignomy before the students of the University of Paris. After him followed the renowned Avicenna, Averroes, Michael Scott, and the Italian sculptor and naturalist, J. Baptista Porta, the discoverer of the camera obscura. Later still, many German, French, English, and American observers left their writings among us to be added to and built upon. Lavater, in 1801, wrote numerous volumes on the subject, copiously illustrated, in which he had the assistance of some of the best artists in Europe. It is through his works, and from his associations, that this science is best known to modern students. His purity of life and high position (he having been an eloquent clergyman, pastor of St. Peter's Church at Zurich) placed Physiognomy on a footing of credibility. His works are what he named them—"Fragments"—merely, without system and largely impractical. His efforts, like those of his predecessors, have assisted in continuing the belief and interest in the science.

Prominent among the German and French observers and writers are the eminent Blumenbach, Spurzheim, Camper, Bichat, Broussais, and De la Sarthe; among the English,

Sir Charles Bell and Alexander Walker; and among Americans, James W. Redfield. In 1817, Dr. John Crosse published from the University Press, at Glasgow, a series of lectures on Physiognomy which he had delivered, setting forth a system which contains practical knowledge, susceptible of proof and capable of application by any ordinary observer. The latest work on Physiognomy, written by Dr. J. Simms, of New York, entitled "Nature's Revelations of Character," is the first that has given to the world a system based upon a natural classification, and corresponding to the other departments of Natural Science. His system, as is this, is based upon form, and, viewed from that standpoint, is unassailable; it should be in the hands of every thinking person. Without pretending to treat this science as exhaustively in regard to principles as Dr. Simms has done, I claim to have discovered and elaborated some laws which seem to have escaped his penetrating mind, and also to have carried it one step beyond his exposition. Professor Joseph Le Conte, of the University of California, in a recent able article in the "Popular Science Monthly," describing the advance of science, says:

"In all sciences, but especially in the higher and more complex departments, there are three distinct stages of advance. The first consists in the observation, collection, and arrangement of facts—Descriptive Science. The second is the reduction of these to formal laws—Formal Science. Thus far the science is independent of all other sciences. The third is the reference of these laws to the more general laws of a more fundamental science—in the hierarchy as their cause—Causal Science. It is this last change only which necessarily follows the order indicated above. Its effect is always to give great impulse to scientific advance; for then only does it take on the highest scientific form, then only does it become one of the hierarchy of sciences, and receive the aid of all. Thus, to illustrate, Tycho Brahe laboriously gathered and collated a vast number of facts concerning planetary motions—Descriptive Astronomy. Kepler reduced these to the three great and beautiful laws known by his name—Formal Astronomy. But it was reserved for Newton, by means of the theory of gravitation, to explain the Keplerian laws by referring them to the more general and more fundamental laws of mechanics as their cause, and thus he became the founder of Physical and Causal Astronomy. In other words, Astronomy was at first a separate science, based on its own facts. Newton connected it with Mechanics, and thus made it one of the hierarchy. From that time, Astronomy advanced with increased rapidity and certainty. Astronomy first rose as a beautiful shaft, unconnected and unsupported, except on its own pedestal. In the meantime, however, another more solid and central shaft had grown up under the hands of many builders; viz., Mechanics. Newton connected the astronomical shaft with the central column of mechanics, and thus formed a more solid basis for a yet higher shaft."

This description truthfully and beautifully shows the progress of scientific research. The system which this work presents to the reader has advanced to the third stage of progression. It presents a description of facts in relation to the human organism which have been observed and collected; it reduces these facts to laws; and, lastly, shows the correspondence of this science to the general and fundamental laws which underlie all matter—viz., those of Chemistry, Architecture, and Mathematics. The sum of all human action is based on these three fundamental principles of Nature, and Man's organism illustrates the influence of these laws. I would like to see the facts contained in this little work in the hands of all who love their kind, and who desire its elevation by scientific methods. In the years to come, I do not doubt that more ample knowledge of Physiognomy will be disseminated by greater minds, with better opportunities of observation than have fallen to me. It would seem a very appropriate time for the spreading of knowledge of Man now that so much is known of his environment, and while so many hitherto unknown applications of the forces and substances of Nature are coming daily to light that are immediately connected with his welfare. Earnest and religious regard for the advance of mankind to grander heights of purity and nobility of life, added to the belief that nothing short of the knowledge of scientific laws and their application can regenerate the human race, has impelled the writing of these ideas.

CHAPTER I.

PRIMITIVE ORIGIN AND DEVELOPMENT OF MAN.

"Looked at through a single science, life is unintelligible; for the sciences, separately taken, are but like the constituent portions of a telescope—we can only see properly by connecting them."—GRINDON.

Tracing Man to his origin, the Monera, or, as some naturalists claim, the Amœba, we find entering into his constitution four essential elements, or primal components. These, in the language of chemistry, are called nitrogen, oxygen, carbon, and hydrogen. These elements are first exhibited in the air and water; they are taken up by plants, and upon these elements all vegetable life subsists. All organized life proceeds from the same elementary powers: First, plant life; then insect organisms; after these, reptiles, fishes, birds, beasts, and, last of all, Man.

In every one of the varied and various organisms the same universal principles prevail; very little of any other elements enter into the composition of any of these bodies. The different phenomena are produced only by difference in proportion, by chemical action and chemical changes. The plants suck up through the roots the nourishment needed to give them form, color, and stability. The leaves also assist in their nourishment by taking up through their innumerable pores, or mouths, the elements which they require from the air. One hundred and twenty thousand of these inhalers have been counted on one small leaf by the aid of the microscope.* Animals feed upon the plants, and the same elements

* Johnston's "Chemistry of Common Life."

still reside in their organisms. The form is changed, it is true, and one might suppose that the animal form was composed of entirely different materials from those which create and nourish the plant; but it is not so. The very same elements are there; there is no addition of other principles; they have simply assumed different forms through the power of chemical action. The same activities which formed the Moneron, or germ-cell, of all organic life continue the process of chemical action, but in a more complex manner, to form the animal organism, as well as all the intervening grades of organic life. Chemical analysis proves this truth. The elements of plant life and animal life are identical, differing only in proportion. These elements take on other forms, and are called by other names, according to the location in which they are found and the organism in which they reside or create. Yet the chemist never loses sight of them, and, no matter what metamorphoses take place, he knows that they are sure to appear, and identifies them regardless of the forms which they assume.

Before these elements can become human organisms, they must pass through many complex and subtile chemical changes. They must have had an existence in minerals, air, and water. After this stage, they pass into plant, insect, and fish life; later on, into the various organized birds and beasts; these, in turn, feed upon and increase by the help of plants and water; and upon plants, beasts, and water, Man is sustained.

This, in brief, is the progress of Man from his primitive creation, as a simple germ or cell. From this starting-point emanates all organized life, coming up through gradational forms until the physical man is reached, and his wondrous mentality exhibited by the aid of his physical phenomena.

What is mind? What are mental operations? These are questions which the wisest minds of all ages have vainly tried to solve. Not until the present epoch have we found the instrumentalities essential to their solution. The microscope, the spectroscope, the telescope, and chemistry, together

with our knowledge of the laws of molecular motion and of
the other forces of Nature, have gone far toward unraveling
what has always been both a wonder and a mystery; and
although the ultimate of knowledge in the direction of mind
and mental action has not been reached, yet we have learned
enough to know that all mental phenomena are dependent
upon the differentiated physical organisms for their demonstration, and that the same four primal elements contribute
directly to the production of all mental action as well as
physical power.   In the beginning of organized matter, these
elements were called oxygen, nitrogen, carbon, and hydrogen; in the blood and tissues of man they are found after
many transmutations, and are then known as fibrine, albu-
men, caseine, and water.   By continued physical and mental
labor on the part of man these elements are exhausted, but
are replenished by a due admixture of animal, vegetable,
and marine foods.   The manner of their use, and how they
conduce to the upbuilding of the human organism, will be
shown in the chapter on Hygiene.

The first use which organized mind makes of its newly
acquired powers which a progressive evolution has bestowed
upon it (as is first shown in animal life) is to commence recreation on a small scale, out of the materials which are
found in abundance suited to the purposes of its peculiar
phase of development.   The manifestations of constructive
energy, as shown by the spider family, for instance, are remarkable.   Here we find no brain proper; it is true there is
a nervous system in simple form.   Its mental energy, for I
cannot name it less, must proceed from the power and intelligence which inheres in this nervous system, for it is shown
that in the human organism nerve and brain matter are
identical; hence, we must conclude that the peculiar form of
nerve matter which is found in organisms that have no brain
proper is the source and seat of intelligence in such bodies,
and that their peculiar forms of mental and constructive energy proceed from the power of their nervous system, as well
as from such organs as are developed in them.   In support

of this theory, let me quote from Prof. Grimes. He says: "The very lowest animal life that can be produced is the amoeba, a mere minute mass of jelly. It cannot be said to have any particular permanent form, but it has the ability to assume almost any imaginable shape, according to circumstances. It can protrude a portion of itself forward in the form of a limb; it can thus produce a dozen limbs; it can spread itself out into a thin sheet, and envelop and absorb what it wants, and then change its form again. This creature occasionally manifests a degree of mechanical skill not surpassed by the beaver, and not even equaled by civilized man. We learn an important lesson here, and that is, that Nature is capable of performing superior mental operations without any special organs that can be perceived. Next, observe the spider. He has nerves, but no cerebrum or cerebellum, no thalamus, no striatum; he has something that appears to be analogous to the human oblongata, and that is the nearest approach to a brain; yet he surpasses the beaver in mechanical skill, the fox in cunning, the monkey in dexterity, and the tiger in malicious cruelty. He surpasses all animals that have brain, excepting man. What is the explanation? The answer is obvious. The spider has the organs of his mental faculties, call them by what name you will, located somewhere in his body, in his nerves, or in his ganglionic masses. He has no brains; no animal has them except fishes, reptiles, birds, and mammals. Fishes have what is called a brain, but it is a mere bud of the anterior lobe of a brain, and that only. There is not in all Nature a more interesting lesson than that which is conveyed by a comparison of the animals that have no proper brains with the fishes (the lowest of those that have them), and then a comparison of these with the next class above them, and then again with the birds that are one step higher still, and then the birds with some of the lowest animals of the next higher class, the mammals, such as porpoises and rabbits; then higher mammals—cats, foxes, dogs, horses, elephants, apes, man."

Dr. M. Foster, an able and recent writer on Physiology,

claims that the amœba is possessed of the following powers and functions: he says that "it is contractile; it is irritable and automatic; it is receptive and assimilative; it is metabolic and secretory; it is respiratory and reproductive." And yet the microscope fails to find a simple nervous system in this animal! The lowest organism in which has been found a simple nervous system is the turbellaria. This organism, according to Haeckel, is the sixth in progressive creation, starting from the monera. It has also organs of secretion (kidneys) and of generation. According to my theory, and the system of progressive growth which I set forth in this work, man represents, in his unfoldment and progress from his lowest development as an embryo, the same order as is observed in the primitive growths or primeval organisms. This system of Physiognomy also finds in the first or lowest division of the human organism the organs of reproduction and the first organs of secretion—viz., the kidneys. The coincidence would be remarkable did not Nature abound in just such corroborations of her unity of action and method. These methods are as potential in the development of man as in the development of the vegetable and animal kingdoms, and outwork in the same manner. "The laws of inheritance and adaptation" are the prime factors in making man what he is. After taking into account man's inherited quality, the remainder of his individuality is the result of his environment; that is to say, the soil upon which he lives and from which he derives his food, the air which he inhales, the water which he drinks, the climate which he dwells in, and the quality of mentality which surrounds him. If the soil upon which is grown his food, his plants, grains, fruits, and cattle, holds in excess certain mineral compounds, then all of his food, as well as himself, will derive their character (aside from their inherited quality) from these sources. He will take his color, his mental powers, his moral and his physical qualities from these surroundings and conditions; and thus man is constantly the subject of manifold laws and processes ever moulding and fashioning his character, his form, color, quality, and all his personality. As the study of the origin and science of color progresses, we shall find that it plays a most important part in our lives and in the history of our planet. Every mineral has its own color; from these mineral compounds the plants, flowers, and fruits get their beautiful dyes. Taken up from the soil, air, and water, these dyes assist in forming character; and by the colors of plants, animals, and man, much of their character is discerned. In the early geologic periods, while yet the whole earth was in a moist and heated condition, the colors of the flora must have been gorgeous in the extreme; but as it approached a cooler condition, the minerals of which the crust of the earth is composed must have parted with some of their coloring power, for we observe that all of the races of plants, animals, and men in the temperate and frigid zones are lighter and have less coloring matter in their composition. As the people in the torrid zone are richer in color, so, as we approach the poles, all Nature is destitute of deep shades. As the earth continues to cool, all Nature, reasoning from analogy, will continue to part with its color, until the inhabitants of our planet will possibly assume the pallor and transparency which is usually ascribed to ghosts and spirits. Never having seen a spirit, I cannot vouch for their complexion, but the prevalent idea is that they are pale and exceedingly diaphanous. Advance in the knowledge of the origin and character of color has been most wonderful in this era. As the intense colors in visible Nature lessened, discoveries of colors hidden in many hitherto unknown substances have been made, and have been applied to remove disease and to the industrial arts in ways never before attempted. The basic color of nearly every mineral has been traced through the medium of chemistry and the spectrum analysis, and the direct influence of the colors of the various plants, flowers, and minerals used in remedies has been demonstrated. This may strike the reader as a singular statement, but why should not color, which is palpable to our sight, be as potent in its influence upon our organism as electricity, the ethers, and other fine forces which are almost imperceptible? What is color for?—to please the eye merely? I have never found anything in Nature that did not serve a two-fold purpose, at least; such is the wonderful economy of Nature.

Every age has its own peculiar form of mental activity, progress, and culmination, corresponding to the growth of man's nature. The Philosophical Age of Greece—the age of Plato, Xenocrates, Polemon, Zeno, and Aristotle—made its impress on the character of the men of that era. Its influence did not end there. All through the Dark Ages, great minds—the Galileos, the Bacons, the Brunos, and others—who could not be crushed through fear of the Church, of superstition, or of death, were stimulated to further philosophical and scientific inquiry by the writings of those old Greek thinkers. Thus, modern Science, following naturally its predecessor, Philosophy, was given to the world. In the order of natural mental progression, inquiry and reason always precede the discovery of laws and principles. Hence, we find during the later years of those "ages," rendered "dark" by the despotism of Ecclesiasticism, a revival of Philosophy, which took a practical direction (the result, mainly, of the activity of Northern minds, aided largely by the inventions which speculation and experiment had produced), and which resulted in the birth of modern Science.

In the days of Phidias and Praxiteles, we had probably the nearest approach to perfection in sculpture that the world will ever witness. In the age of Rubens, Titian, Michael Angelo, and Guido-Reni, the height of configuration and coloring as an art, purely, was reached. The Middle or Dark Ages gave us the acme of superstition and religious fervor, which have produced a natural reaction in the present epoch of reality and science. This has achieved a marvelous number of discoveries and practical inventions.

We are, as yet, upon the threshold; the culmination in this direction has not been reached. The wonders which coming years may disclose can neither be imagined nor depicted. And thus we see that Evolution brings first one side and then another of Man's wonderful mentality face to face with the resources which have been hidden away in Nature's great storehouse, awaiting, seemingly, the time when Man's progressed nature shall demand them.

CHAPTER II.

BASIC PRINCIPLES OF SCIENTIFIC PHYSIOGNOMY.

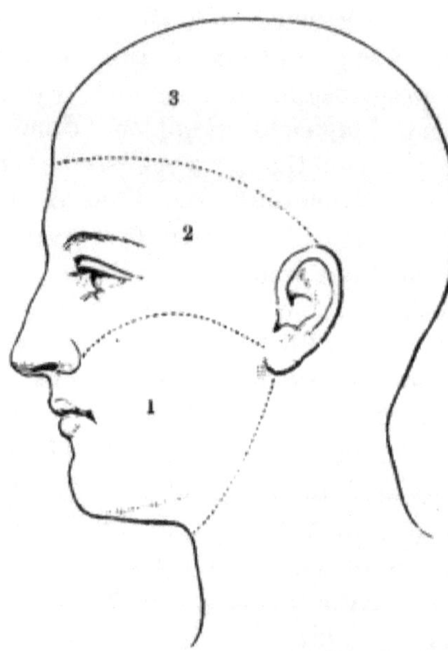

THE THREE GRAND DIVISIONS OF THE FACE.
1, Chemical. 2, Architectural. 3, Mathematical.

Standing at the apex of all creation is Man, the very epitome, sublimification, and essence of creative energy; what more natural than that in this high and complex organization should be found, in combination, all of the components of what may be termed the lower creations?

Man is literally made up of the "dust of the earth;" considered as a chemical compound, in Man will be found, made into solid bone, muscle, brain, blood, and tissue, not only the "dust of the earth," but also nearly all the elements contained in the earth. In his composition will be found oxygen, nitrogen, carbon, hydrogen, calcium, iron, sodium, chlorine, sulphur, phosphorus, potassium, and a small amount of other minerals.

In the face of Man will be found, by dividing it into three grand divisions, the signs of character representing the three basilar principles underlying all matter, as well as Man's own organism; viz., Chemistry, Architecture, and Mathematics.

If one examines a grain of sand, and enters into an analysis of its constituents, he finds that it has, first, chemical properties—so much of one kind of element, another particle of some other sort; perhaps several other elements enter into its constitution. These various elements have an affinity for each other, and harmonize in their combination; this is the power which binds them in one, and forms them into a *chemical compound*.

Upon further examination, it will be found to possess a definite *form*. In the case of crystals of the various minerals, this form is always defined by *law*, and the mineralogist recognizes each object by its *form*. This natural Law of Shaping of all objects, both animate and inanimate, is an illustration of *Architectural Law*.

If the crystals be reduced to their elementary particles, the *number* of their constituents is discovered. This is the *Mathematical Law* exemplified.

All creations, from a grain of sand up to the planetary bodies, have their Chemical properties, their Architectural *formation* or shape, and the *number* of quantitative particles which Mathematical Law requires.

The same constituents which compose planets, which form minerals as well as plant, insect, and animal life, form also Man's organism. These elementary constituents bring with them, into Man's body, their basic principles; and wherever we find Man, we can but observe that in the chemical action of the elements composing his body and surrounding him— that in his *form* and *proportions*, and in the *number* of elements entering into his constitution—the same laws of Chemical action, of Architectural formation, and of Mathematical quantities or particles, which govern all other departments of life, are as potential in fashioning him and in determining his character.

It is thus shown that man, in himself, in his own person, typifies all creation; proving that he is the very essence, the subtile, refined organization or force evolved from all forces, powers, causes, and chemical actions in the universe. He is, in short, the highest creation of which we have knowledge.

A correct understanding of this grand organization is the first science in the world; the first in importance to each one of us. It has its laws, which are exact and yet complex; but where is the reader skillful enough to understand them? As Nature is perfect in her works, and has made few laws so mysterious as not to be comprehended, is it not natural, then, to suppose that man is capable of understanding his own organization and the laws which govern it? He may, if he will but seek the truth and fear not.

As the dial is to the clock, so is the human face to Man; it is his exponent, morally, mentally, physically; on it are written not only his mental powers, his moral strength or weakness, but also his physical capacities, powers, weaknesses, and predispositions to health and disease; and there is no one of ordinary capacity who cannot learn all those signs almost at a glance. The importance of this knowledge is incalculable. Inasmuch as we all have to pass our days in intercourse with our fellows, it is of the greatest importance, not only that we should understand ourselves, but also that we should be able to comprehend, to a nicety, all with whom we associate; not merely for our protection and the pleasure we may derive from it, but also for the good that we may do; because this knowledge will teach us that what we now call "charity," in overlooking the faults and weaknesses of others, is but simple justice, for it is not just to expect more of an organization than Nature has given it power to accomplish. Therefore, we may spare our "charity," and through knowledge give *justice*.

The three grand divisions of the face—namely, the Chemical, the Architectural, and the Mathematical—have also their subdivisions. The Chemical includes the moral, the domestic, and the supplyant powers; the Architectural, the faculties which indicate the building, artistic, religious, and literary traits; and the Mathematical includes the reasoning powers, which are the chief faculties in numerical demonstration.

We shall proceed now to the examination of the subbasilar laws of the science. The first is stated thus: All form indicates character. The second, all color denotes character; third, texture is significant of quality; fourth, the size of the nose, *controlled by quality*, is the measure of power; fifth, the shape of the nose shows the kind of power; sixth, Nature is harmonious; seventh, Nature is compensatory.

These last two propositions are established throughout all Nature's works. They are proved in all the lower animals, among birds, insects, and plants. Observe, for instance, the spider, which has no wings to assist it in escaping; its compensation consists in having eight eyes, placed two in front, two on top of the head, and two on each side. These eyes are without motion; yet their situation enables the spider to comprehend every view which his safety demands.

We now commence with the general proposition that "all form indicates character." Within the three grand divisions of the face, we find the facial indications of five different systems of functions which go to create the different forms of man, and which are always found in combination, but in different degrees of development in different persons. These are named the Vegetative, the Thoracic, the Muscular, the Osseous, and the Brain and Nerve systems. Upon the different degrees of development of these several conformations depends man's power for being mainly either Chemical, Architectural, or Mathematical.

The organization which is mainly Chemical in its operation and effects is known by a predominance of the Vegetative system, and is accompanied most largely by all those functions which serve to supply the body with material, and for the protection and procreation of the race. The functions included in this division of the organism are those of digestion, reproduction, respiration through the mouth, secretion, excretion, and growth. These functions are productive of the following faculties: Conscientiousness, Firmness, Alimentiveness, Benevolence, Amativeness, Love of. Children, Mirthfulness, Approbativeness, Modesty, Self-esteem, Secretiveness, Resistance, and Force. These include in their action all the laws common to vegetable life, and have not the power of motion, will, and conscious sensation—these being exclusively animal powers. The development of all these traits proceeds mainly from chemical action; as, for instance, the sustentation of the body and the procreation of the race. These operations are entirely chemical.

The Architectural division is shown by a predominance of the Muscular, Thoracic, and Osseous systems, which embrace within their own action almost all of the principles of mechanical forces; such as the different lever powers, different principles of valves, and the representation of a pulley (in the action of the superior oblique muscle in rotating the eye); also other powers which will be mentioned hereafter. The traits indicated in this division are Hope, Cautiousness, Analysis, Imitation, Ideality, Sublimity, Human Nature, Constructiveness, Acquisitiveness, Veneration, Self-will, Credenciveness, Observation, Memory of Events, Form, Size, Weight, Color, Order, Calculation, Locality, Music, Language. You will observe, by these names, that the artistic and religious faculties are included in this as subdivisions.

The Mathematical division of the face has its work performed mainly by the Brain and Nerve system. The faculties shown in this division are named Time, Causality, Comparison, and Intuition.

The several systems of the body and faculties of the mind act and react upon each other, and sustain inter-relations to each other; but each division is mainly sustained by the action of the system to which the several different parts of the face indicate it as belonging.

As I have before stated, the principles of Physiognomy are founded on the same general laws which underlie all matter, but having for their demonstration special laws. When we reflect that brain, matter in the form of nerves, and nervous ganglia, as well as the muscles, are instrumental in producing mental manifestations, we must at once conclude that the rather contracted views and theories of the ancient metaphysicians and modern phrenologists must give way to more extended and well demonstrated *facts*. The entire surface of the body being covered with a cuticle, upon which a fine net-work of nerves ramifies, gives us a very extensive sense-organ, and makes us cognizant of temperature, tactile sensations, and pressure; and by the aid of these several sensations, very many *mental* impressions are conveyed.

The theory of Mind which is set forth in this system of Physiognomy is more comprehensive than any which has been given hitherto. Many advanced and eminent scientists and physicians to the insane have recently become imbued with the idea that the brain is not the *sole* and *exclusive* mental organ; that the muscles, and the nervous ganglia and plexuses of human and animal organisms, *may* be of a mental character, and exhibit or assist in illustrating mental manifestations.

Among those who advocate this idea as probable, I may mention Herbert Spencer, Dr. Maudsley, and Dr. Lindsay—men whose opinions are received with respect and credence. It has been reserved for a woman, however, to carry their observations and research to a finality; and by the aid of the physiognomies of both animal and human beings, their ideas of the diffusive *locale* of the mind have been extended and made more comprehensive still, by *proving* that the *viscera*, as well as the bones, muscles, and nervous ganglia and plexuses, are instrumental in exhibiting mental phenomena. It is not essential, at this stage of this work, to give proofs of the position I assume in this theory of Mind; but as we proceed, the *rationale* will be developed, and I believe the evidence will not be wanting to substantiate my position.

I leave the case in the hands of the scientific, the logical, and unprejudiced reader. My motive is based on a love of Nature and of justice, and will enable me to reject any idea, however much I may respect it, if it be found not true and scientific.

That Physiognomy, as a science of the *mind* and *body*, has been understood in some remote ages of the world, is proved by reference to Leviticus, xxi, 18–21. You will find that that great law-giver and hygienist, Moses, understood what I claim for Physiognomy to-day; viz., that all personal defects which are congenital, and not accidental, are the outward signs of mental and moral deficiencies; and he demanded that those who came before the people to be their spiritual and moral guides should illustrate in their own persons that physical combination which is the indication of moral balance and intellectual capacity. In his directions to those priests who were to serve at the Holy Table, he says: "For whatsoever man he be that hath a blemish, he shall not approach. A blind man or a lame, or he that hath a flat nose or anything superfluous, or a man that is broken-footed or broken-handed, or a dwarf or crooked-backed, or that hath a blemish in his eye, or be scurvy or scabbed, or whatsoever man he be that hath a blemish, let him not approach." Indeed, the whole tendency of Moses's teaching proves him to have possessed a knowledge of sanitary law and mental characteristics such as Scientific Physiognomy only teaches, and which has made his followers the longest lived, the healthiest, and most prosperous people on earth to-day.

The index of both body and mind is written in unmistakable characters on the face; and it is most astonishing that while people have a knowledge of almost every other science, that of the face is a sealed book to them. The mass of the people are, through ignorance of natural law, generated under all sorts of horrible conditions. They perceive results only—not causes; and to erroneous and false systems of religion and metaphysics are we indebted for this lamentable condition of mankind. Not until we understand natural laws and apply them—particularly those of Physiognomy and Physiology—shall we have higher types of manhood.

The basilar principles before mentioned, discovered by me and verified by a lifetime of observation, investigation, and reflection, I give to the world as my contribution toward the enlightenment of its denizens. I do not claim to have discovered all that exists on the subject. As an eminent geologist has told us, "Man is not made, but making," so the knowledge of man must go back into the past and keep pace with the present, to know what may be the possibilities of the race in the future. One person can no more tell us the all of human character than can one astronomer tell us the all of the great worlds above us.

The ancient philosophers believed the mind to be a unit, operating independently of the body; Phrenology went farther, and showed some of the relations of the mind to the brain; Physiognomy goes still farther, and proves the relation and interaction of the faculties of the mind and brain with the several organs of the body, and locates their position in the face exactly. As we proceed to study this science, we must keep this fact ever before us: That faculties or traits are dual, and have their physical, as well as mental representatives; that neither can operate without the other. The face proves this; the life of man bears witness to it. There is no line of demarkation between the different parts of our natures, separating one from another, and all mental, moral, mechanical, artistic, and literary phenomena find their sustentation and illustration in the numerous physical powers which make up man's organism.

Scientific Physiognomy, well understood, declares to us that the human body and mind are regulated by a system of *checks* and *balances;* as, for instance, where we observe Self-esteem lacking, we always find some other trait in excess to supply the want—to assist, as it were, in balancing the character. Just so, in the physical organism, where one organ is weak or defective Nature at once gives assistance by calling upon some other part of the system to supply the deficiency. Thus, to illustrate, when the kidneys are small or defective the skin is unusually active and assists in carrying off the waste of the body by means of perspiration. In the same way a great excess of one faculty shows a corresponding weakness in another. This applies equally to the physical as to the mental. Whenever you discover one great trait in a character, whether it be a fault or a virtue, endeavor to ascertain for what it is the compensation. Lord Byron and Edgar A. Poe are illustrations of this principle. Each possessed a predominance of the mental faculties at the expense of the moral, and their physical organisms were correspondingly unbalanced. Both died young.

Horace Greeley is another instance of the truth of this proposition. He was greatly lacking in several practical faculties; his compensation consisted in an excessive and uncommon development of Reason and Memory. His physical powers partook of the same strength and weakness, and when a time of unusual exertion came to him the inequality of the mental and physical powers became apparent; his mind gave way, and this great and brilliant man died in a mad-house.

The judgment of the masses as to what constitutes greatness is usually very superficial. Their estimate of a great man is that he is universally great—great in every direction; while the truth is that the man who shines so brilliantly in a given sphere is usually very much wanting in some part of his organism. Physiognomy will discover the deficiences and locate them exactly in the face. They can be seen at a glance almost, and any one of ordinary comprehension, with a few simple instructions, can discover this readily; this test is infallible. Were this knowledge more general it would benefit mankind incalculably; as, for instance, in selecting the right partner for marriage, in choosing friends, and in deciding the natural adaptability to employments, and in comprehending the mental and physical constitutions and conditions of children. To mothers, it would give the power to understand, almost instantly, which organs of their children were weak and which strong. One can locate as readily the faculties in the face, after few instructions, as one can point out the rivers, bays, and mountains on a map; it is simply Human Geography. At every step of life it is of use. At home, at church, in the street and in society, one can make studies with trifling effort. In the whole range of the natural sciences, I do not know of one so beautiful or more profitable than this. It is elevating to the intellect and moral character, and nothing in Nature presents greater proof of the power and wisdom of God than this science. In short, it is the *duty* of every one to understand himself and herself, as well as those by whom we are surrounded. Many persons, on finding that the face discloses character, are seized with fear. Let me ask such, for what do they suppose the face was made? Perhaps for the same purpose for which Talleyrand said words were invented—to conceal thoughts. Not so!—but to reveal them. God did not place us here to live forever under laws which were to be a sealed book to us, but He gave us the faculties to discover and apply them.

In commencing the analysis of character, all self-love must be thrown aside, and the individual must be willing to stand before the whole world for just what he is according to the rules of science—a correct knowledge of which will give a true analysis of character and will also teach the methods by which knowledge comes to us; through which faculties we receive an understanding of Size, Form, Events, and Numbers, and which department of our Memory is faulty, and how to strengthen it.

Very few persons analyze keenly enough to discover through which faculty they learn spelling, geography, marksmanship, and other arts; also, most persons have very erroneous conceptions of the *rationale* or philosophy of the emotions and passions—love, jealousy, revenge, benevolence, generosity, hatred, and the like.

One will hear many persons declare that there can be no real love without jealousy. An analysis of any character in which jealousy preponderates will give a character very unbalanced, and lacking either in its moral or mental make-up. "Perfect love casteth out fear." Suspicion, jealousy, hatred, and revenge are always the accompaniments of unbalanced natures, and are the proofs of a lack of either good reasoning powers, conscientiousness, self-esteem, or perception—one or more; and when we perceive either of these passions exhibited, we will find in the face the proof of this statement disclosed.

The why and wherefore of all the defects of human nature never has been understood, through the want of a system to analyze and verify the indications. In the same way, many traits are misunderstood and their action not comprehended, for lack of scientific knowledge of character. Thus, for example, the faculty named Amativeness, or Love, which is, we may say, the power underlying all the other faculties, and really one of the most important in the human economy, is the least understood of all the faculties. Physiognomy proves this to have a *mental power*, as well as a physiological basis, and to be possessed in a larger degree by all persons of *creative talent* than by others not thus endowed. This creative ability proceeds from an excess or a large endowment of the *procreative* or reproductive system. This principle will be understood better as the reader advances; the plan of this work is progressive, leading from the simple to the complex. But let me state, just here, that it is an established law of Physiognomy, that a person can illustrate best in his works those principles and laws of Mechanism, Art, and Science which are found well developed in his own organism. It will be found, upon further study of this system of Mental Physiology, that it is the procreative power which guides the graver's chisel, the artist's brush, the composer's pen; and is illustrated in the musician's harmonies, the imagination of the writer of fiction, the inventor's mechanism, and the dramatic artist's imitations. The *proof* of this can be established by the study of the faces of Mozart, Rubens, Michael Angelo, Shakspeare, Dickens, and all persons in all ages who have excelled in any department of

Creative Art. They could not have thus excelled without this function and faculty preponderating.

The beautiful child-characters conceived in the brain and portrayed by the pen of Dickens are creations as real as though the breath of life had been breathed upon them. They could not have had a mental conception and birth, had there not been a large endowment of the procreative power in the organism of this author. The face of Dickens proves him to have been possessed of the faculties of Amativeness, Love of Young, Constructiveness, Human Nature, Form, Size, and Imitation, in a remarkable degree; and one who has learned to localize the signs for these faculties can find the evidence of them in the physiognomy of this great character-painter.

Physiognomy teaches the laws of Heredity, and shows how ancestral types are reproduced; it teaches how, by the right mingling of types, to eradicate weakness, both mental and physical, as well as vice and immorality.

In order to bring about a higher humanity, we must reverse the methods and opinions which have prevailed for ages, and, instead of regarding these subjects as bad and degrading, we must teach that religion commences in the physical system, and that the surest way to save sinners is by learning the grand truths which Science unfolds, and which must be applied in order to bring rightly organized bodies and souls into existence.

A celebrated writer on Heredity says: "As yet, there has been no true breed of men under the accidental or intuitional action of love. There is a superabundance of imperfect men and women on the globe, with only here and there a specimen that suggests the possibilities of the race." We must bring our eyes close down to Nature, and there learn the lessons which never can be false, not depending for guidance on doctrines and theories, venerable though they be, which have been tried and found utterly wanting in every element of true science and true religion, "which is the fulfilling of the Law."

I maintain that nearly all the errors in regard to Man—his life, his surroundings, his relations to them and their relations to him, his religion, his sense of right, his misconceptions of beauty, his exceedingly scant knowledge of governmental principles—proceed directly from *utter ignorance of himself*, and while he has a knowledge of the planets, stars, winds, rocks, beasts, birds, snakes, and animalculæ, he does not know the laws which govern his own body. He cannot understand one single sign of character as indicated by the face; he knows not the meaning of different voices; the walk of man conveys to him no meaning; the color of the eyes and hair declares nothing to his sense of sight. He is like a mole, groping in daylight. He plans and executes grand enterprises; he spans continents; he examines the character of the uttermost stars, calculates eclipses, traces the paths of comets to remote ages, understands to a nicety the great world and the little world as shown by the telescope and the microscope; and yet cannot sound the depths of his child's character, which appear to him unfathomable. Why is this? Is it because the science of Man is more abstruse and occult than all others?—because it belongs to the unknowable? Not so. It is because he has not thought of these things, and because he has not been taught them as he has the other sciences. I regard it as the most simple of all the sciences, the most easily demonstrated, the most essential to human happiness and welfare.

And until the science of Physiognomy is commonly understood, Government, as a science, cannot go forward. Legislating for beings, of the laws of whose existence one is in utter ignorance, is an absurdity, and will fail. Not until the masses can put themselves in harmonious relations to their environment can government go forward, and this can result only from a complete knowledge of Man, his capacities, his needs, and his possibilities. This knowledge proceeds only from a scientific study of himself. When Man becomes convinced that his face registers his life, and that "he who runs may read" what he has been about, and that he cannot hide his inner self from the gaze of the world, he will endeavor to make his life so good and so noble that he will not be ashamed of the most rigid scrutiny, because it is only in thus doing that he will be enabled to have either a character or a reputation.

Physiognomy as a science, with rules and established principles so plainly set forth as to be comprehended by the masses, has never been given to the world until recently. Lavater possessed the power of reading the human face intuitively, but he has left among his writings no rules nor principles by which students can learn this science. The best book and school for students is Nature. Still, a keen observer may record such discoveries in this field as to be a benefit to coming generations. This science is gigantic in its proportions; and when we reflect that there are in the world no two organizations with exactly the same combination of traits, we see that the field is wide, with room for many observers.

CHAPTER III.

THE FIVE SYSTEMS OF FUNCTIONS THAT CREATE CHARACTER.

"The mind is invisible to those who understand not the body of Physiognomy."
WINKLEMAN.

Victor Cousin, in his admirable "Essay on the Beautiful," remarks: "All is symbolic in Nature. Form is not form only; it unfolds something inward." This philosophy is scientifically correct. That all form indicates character is a principle so well established throughout Nature as to need little testimony from me. In the study of the science before us, it is absolutely essential that this principle should be thoroughly comprehended, and the character of its various phases understood at first sight, in order to render the interpretation of character certain and beyond all peradventure and doubt, for upon the conformation of the Physiognomy (and here I mean the entire body) are we mainly dependent for knowledge of the character of the entire man. It is true that size, color, texture of the skin and hair, health, etc., play their part in determining characteristics; still, Form is primarily the grand determining, dominating principle underlying all others. Its meanings should be completely mastered before proceeding to the consideration of other branches of our subject.

The more advanced Phrenologists, who commenced the investigation of Phrenology on the basis of classification by *color* of the complexion, hair, and eyes, have gradually arrived at the conclusion that *form* is the most decisive factor in the interpretation of character. O. S. Fowler declares, in his work on "Human Science," that the correct way is to classify character by the forms of the body, and that these forms are produced by the predominance of one or the other of the five principal systems of functions included in the human organism. These systems and forms he designates the Vital, Motive or Mechanical, and Mental Temperaments.

I cannot comprehend how Mr. Fowler can consistently retain the word "Temperament" in his designation of *forms*. Temperament is the word used by Hippocrates to indicate the several colors of the complexion. It has no more relation to Form than it has to Color. If we wish to use language at once intelligent and comprehensive, it must be rejected, as well as the method of deciding character by so small a portion of the organism as the skull alone. Why should not the face, where the most active and impressible of the muscles are located, and the contour of the entire body, be taken into account? It is certainly a good deal more difficult to feel the head, which has no power of expression, and is not as practicable for every-day and instant use as the face and outlines of the body. "A cat may look upon a king," and so one may study the features of his fellow-men without saying, "By your leave." This system of Physiognomy classifies upon the basis of *forms*, which are derived from the several organ-systems that are included in the human body. It also shows the influence of color and health, as well as all the other conditions which determine character.

In the human organism there are five different organ-systems, which, in their development, produce different configurations of the body and corresponding differences of mental development. These systems are always found in combination, but in different degrees of power, in every individual, and to these variations are we indebted for the infinite variety of the human race. These five systems are named the Vegetative, the Thoracic, the Muscular, the Osseous or Bony, and the Brain and Nervous systems—the last mentioned forming one system and producing one conformation. Each of these systems exhibits a different set of physical functions and mental faculties peculiar to itself, but are so constituted that neither can exist without the action and interaction of some proportion of each of the others.

In order to create a normal and healthful condition of the organism, there must always be a due development of each of these systems; else incompetency, disease, and early decline will be the result. The system which is the first in the order of development of all organized life, and which is also the underlying or basilar system of man's organism, is the Vegetative, and has, in common with the various growths of vegetable life, the functions of sustentation, imbibition from the air and water (through the pores of plants and the mouth of man), the functions of reproduction, of assimilation, of absorption, secretion, excretion, respiration, circulation, and growth; but the Vegetative system gives no power for the expression of the phenomena of either thought, volition, or sensation. Every plant, tree, vegetable, and shrub has the power of absorbing, excreting, reproducing, circulating its sap and juices through its cells and tubes (and this by hydrostatic law and the law of gravitation). All the lower animals have the same powers and functions. Not until other systems of functions are superadded do we discover any power of volition or conscious sensation. It is true that the lowest animal organisms give indications of possessing a certain form of sensation, yet these are all in the *vegetative condition*. No organs for the expression of sensation, as we find it illustrated in higher animal organisms, have yet been evolved, and until these organs or systems of functions are added—such, for example, as bones, muscles, and brain— volition, sensation, and thought, in their most complete sense, are not present. The intelligent reader, who has followed the course of the evolution of man from the lower organisms, will have observed the order in which the several systems of functions, and their accompanying faculties of Mind (as it is called), have evolved or developed. As Nature has indicated this order as her supreme law of progression, I shall endeavor, in the exposition of this system of Physiognomy to follow her methods, believing the laws of Nature to be divine and infallible. These laws, if allowed free scope and not impeded by the ignorance or willfulness of man, will always result in harmonious development and equilibrium.

A study of the laws of natural progression shows us that all organic life commences with the development of the functions of sustentation, reproduction, secretion, and excretion. Here, then, is the physical basis of life. "Man, in his embryonic life, passes through all the various stages of progress and development that the lower organisms pass through in their evolution from the merely vegetative existence to the highest degree of sensation attained by animal organisms."* At his birth he possesses all of the functions and faculties which characterize all vegetable and animal organisms, with the addition of a higher grade of intellectual apprehension and with more perfected and sensitive members and faculties. *These are arranged in the body in the exact order in which they have evolved in the lower organisms*, the vegetative functions occupying the lower portion of the trunk, and the brain, the latest organ to become developed, the highest portion of the

* Haeckel.

body. *In the human face, the signs indicating these several organs and functions, with their accompanying faculties, will be found to follow the same plan precisely.*

THE VEGETATIVE SYSTEM.

The amœba, the first and lowest specimen of animal life, like Topsy, "never was born. It growed." At its starting-point it is a mere speck of jelly, without form, swayed by the waters in which it exists into any shape the waves compel it. It appears in this stage to be nothing more than stomach—stomach all over. It lives by assimilating nutriment which it derives from the air and water. It is expressionless, shapeless. It is now simply a *chemical compound.* The rudimentary stage of all organic life is chemical merely. After a while this speck of jelly, in its next stage of progress, begins to attract from the sea particles of lime, and shape for itself an outer covering; this is its bony system. At this point of growth the bone is placed outside instead of inside the animal. It has now taken upon itself the next step in evolution, and become architectural as well as chemical, and assumes various fixed geometrical forms, as we find in the foraminifera. It is probable that the first races of men were stomach men merely; that is to say, they lived in the lowest range of functions and faculties —viz., those of sustentation and reproduction—and that the powers which assist Man in his Architectural and Mathematical efforts were not developed at that stage of evolution to any appreciable extent, but were merely rudimentary.

The New Zealander is a fair sample of this stage of evolution. Of course, the germs of *all* the five systems were present in the lowest types of man, but in that epoch of development they lacked both size and quality, and were not perfected as at the present time in perfected races. The other systems of functions, and many faculties now seen in combination with the Vegetative, have been very largely perfected since—particularly the Brain and Nerve system, which is now in a more active state than ever before in the world's history. From being a stomach race we are becoming a brain race. What evolution will do for us in the ages to come, it is difficult to predict. The law of progress is always from the lower to the higher, and surely we can say of human nature there is need of higher growth.

The Vegetative system will always play an important part in the human economy. It is the base of many fine traits of character, as well as being the sustaining power of the organism. The absence of a due admixture of this system produces an impoverished body and a mind devoid of many beautiful and useful traits. Friendship, Approbativeness, Conscientiousness, Parental Love, and Amativeness are directly related to this system, and all sustained by its development and action. The Vegetative system is known by a preponderance of soft fatty tissue all over the body, fullness of cheeks, large mouth, slow motions and weak mental action, slow pulse, and face with little expression. People in whom this system predominates are never deep thinkers, are unexcitable, sensual, wanting in dignity and enterprise, generally domestic and social. This system constitutes the chemical or underlying basis of the human organism. It is almost entirely chemical in its operation, and sustains those faculties which also lead to chemical action; as, for example, the procreation of the race, love of children, friendship, etc. The facial indication of this system is shown more particularly developed in the part of the face below the nose. A line drawn across the face below the base of the nose, across the cheek, up to the bottom of the ear, will show in all persons the degree of development of the Vegetative system, although where this system predominates the entire face, as well as the body, will indicate its supremacy. The Vegetative system includes all the viscera within the pelvic basin. These are named the intestines, spleen, bladder, kidneys, pancreas, and organs of reproduction. The action of these organs leads directly to the development of the following named faculties: Conscientiousness, Amativeness, Alimentiveness, Love of Young, Benevolence, Mirthfulness, Approbativeness, Friendship, Modesty, and Firmness. The diseases which affect this system are dropsy, inflammatory rheumatism, gout, apoplexy, tumorous growths, and various chronic disorders.

The Vegetative system shows the functions of body and faculties of mind that are dominant in childhood; and the Physiognomy of all infants and young children, if they are healthy, exhibits a larger development of the signs of those faculties and functions than of the other divisions, where the signs of the Artistic, Mechanical, and Mathematical are located. The three divisions of the face, by the very order of their arrangement and locality, indicate the order of progressive growth and development of the body and mind: First, the Vegetative system, which is dominant in childhood; afterward, the Mechanical faculties assert their dawning powers, just in the ratio that the bones and muscles strengthen; and when children commence to understand the use and management of material substances, they can become expert in light manual labor, both artistic and mechanical. Later in life, the brain becomes more mature, and pure abstract reason dominates all the previous developments, and mathematical calculation serves also to assist mental processes.

The faculty of Conscientiousness is located in the Chemical or underlying system, proving conclusively that morality commences in the physical basis—just where it should, to be of the greatest value to the organism. If a cultivated and quickened moral sense were brought to bear upon all the faculties and functions of this division of the human organism, many of the sins and evils affecting the human family would be unknown. A judicious mingling of this

4

system with a suitable proportion of the other systems of the organism creates health and happiness. How to accomplish this result will be shown as we proceed to investigate the science of Physiognomy. This system constitutes the Chemical division of the body and mind; the faculties set in action by its laws are mainly chemical in their operation.

THE THORACIC SYSTEM.

THOMAS H. BENTON.

We are taught that in the early geologic periods, the development of the organ sytems of the primary vertebrates was in the following order:

First, the intestines; second, the swim-bladder.

And this corresponds to the true Lung or Thoracic system in later organisms. The muscles were the next evolved; then the spinal axis and the bones; last of all the brain proper made its appearance. In the fish, the earliest vertebrate, the lung is rudimental, and is known as the "swim-bladder."

And in the fish we have the basilar plan of all vertebrates; and as Man is the highest development of that class, I shall follow the same line of unfoldment in describing his formation, believing that the methods of nature create a unity of action and universality of type, and when we wish to understand principles concerning the human organism which seem obscure or mysterious, a practical "interrogation of Nature" is the readiest way to solve the problem. Nature never errs—never lies; it is the only "infallible" power of which I have knowledge. "The breath is the life thereof;" without a fair proportion of the Thoracic system (which includes the lungs, liver, and heart), Man would be inefficient, short-lived, and groveling. This system, in its highest manifestation, shows that the aeration of the blood is performed on a large scale. This induces buoyancy of spirits, quickness and clearness of the brain, ambition, hope, and progressive mentality. Pioneers, discoverers, warriors, ambitious and aggressive people the world over, will be found within this class. History abounds in the record of men with large lungs and small brain, who have made of life a brilliant success, but is almost void of those possessed of large brain and small lungs. This system is recognized by large chest, wide cheek-bones, bright, hopeful eyes, elastic step, large nose, wide nostrils, good complexion, happy disposition, not given to intense study, with moderate-sized brain and abdomen. Persons in this class are fond of amusements and out-door sports and business in the open air; quick at apprehending everything, and as quick to drop a pursuit. This peculiarity causes them to excel in pioneering and geographical discovery, and in all pursuits where activity and constant motion are needed. The diseases to which this system is most liable are acute and inflammatory. All the great warriors and orators of the world have possessed a large share of this system. Julius Cæsar, Wellington, Bonaparte, and Cromwell, among warriors, and Cicero, Patrick Henry, William Pitt, and Henry Clay, among orators, are illustrations of this conformation well developed.

The Thoracic system is included in the Architectural division of the face and body, and assists, with lung power and activity of the heart and liver, in the promotion of the many and varied activities which mechanism calls into play.

Hope, Cautiousness, Human Nature, Sublimity, Analysis, and Imitation are directly related to and governed by this system. Persons with this system dominant are characterized by their ready apprehension of natural laws and principles, and appear to be in accord with the laws of Nature; yield readily to hygienic treatment when attacked with disease; and, as their ailments are generally acute, must have prompt attention and relief. They are very susceptible to atmospheric influences, and are easily affected by the poisonous gases engendered by crowded assemblies; indeed, bad air affects them sooner than it does those with small lungs, for the reason that as they inhale *more air* in the same time, the entire organism becomes permeated with the noxious effluvia arising from a crowded hall or theater. Children in whom are found the salient points of the Thoracic system are restless, eager for change, quick to learn by perception rather than from books; are seldom profound scholars; never apply themselves in childhood, but as they advance in life are more capable of continuous application; are adapted only to professions which are pursued mainly out of doors; the study of the natural sciences, and the pursuit of science professionally, is best adapted to this class. The faculties of mind to which their physical functions give rise endow them with just the combination of traits that is essential to success in this department of knowledge. Such should become botanists, stock-breeders, floriculturists, geologists, hygienists, or follow similar pursuits. By this method, their health, happiness, and usefulness will be enhanced.

People of this type make cheerful and safe companions, for, as their organisms are filled with the oxygen and ozone of the atmosphere, their *moral sense* and purity of mind are stronger than in weaker developments of the Thoracic system. They retain their youthful purity, spirits, and vivacity to an advanced age; also, they are, as a rule, high-minded, filled with noble and philanthropic desires, or ambitious to fill prominent and distinguished positions in society.

Thus much will capacity for and a full supply of pure air

do for the individual. A system of ventilation for public buildings and homes is the greatest necessity of the present age. We cannot expect pure-minded, noble characters to thrive and expand in close, ill-smelling, noxious dwellings. If we desire moral men and women, and those who are truly religious, our systems of drainage, sewerage, ventilation, and water supply will have to be amended before such result can be secured; for any system of Theology or Ethics, which does not include Natural Law as its ruling principle, will create no improved types of the human family, and will only succeed in producing a class of theoretic sentimentalists, *without the power* to be either *pure-minded,* noble, or *truly religious.* Fresh air, pure water, bathing, hygienic diet, and self-control, used according to law, contain all the fundamental principles of true religion, and will advance civilization to grander heights of purity, morality, and truth than all the dogmatic theologies of the centuries. Pure water and pure air are the first necessities of life, and must be had if a fine development of the Thoracic system is desired.

THE MUSCULAR SYSTEM.

Great size is not, as some imagine, always an indication of great strength; it may be, and often is, an evidence of weakness. A person with a large, bony frame, without a due proportion of muscle, is found to be very lazy; he dislikes motion, and is not so

JAMES McFADDEN, A NOTED BURGLAR.

strong as his appearance would indicate. On the other hand, an individual possessed of a large Vegetative development is not strong because the Bony and Muscular systems are not sufficiently powerful to carry around the large quantity of soft, fatty tissue which this system exhibits. The result in either case is weakness. Beginners in the study of the science of Physiognomy find it somewhat difficult to distinguish between fat and muscle—between the Vegetative and the Muscular developments. The signs of this system are located all over the body, from the eyes to the tips of the fingers and the ends of the toes.

It is not my intention to go into an extended description of the Muscular system; this can be obtained from any work on Anatomy; but a few simple instructions, to enable the observer to distinguish the salient features of the system, is all that will be needed in this connection. A broad form, with well developed muscles and tendons, quick, elastic step, shoulders broad in proportion to the rest of the body, low forehead, and relatively small head, *large, full convex eye,* short, thick nose, and short, thick neck, are the chief indices of the predominance of the Muscular system. Mentally muscular people are seldom gifted, but under the excitement or stimulus of the emotions, will succeed in many difficult enterprises; under the influence of rage, will become desperate and destructive, and should guard against the exercise of this passion, as really well meaning persons have committed capital crimes by losing control of their muscular powers.

The many and varied expressions of the human face are due to the action of a great number of muscles, some anatomists describing as many as thirty-six pairs and two single muscles in the face alone, and in the entire body more than five hundred. All eminent actors, singers, musicians, and sculptors are largely endowed with a fine quality of muscle. The nature of their professions requires that they should be able to have *perfect control of the muscular sense.* A scientific reading of the faces of all who excel in any of these pursuits shows that the Muscular system is pre-eminent in their organisms, with a suitable Brain system added.

Part of the muscles are voluntary and under the control of

the will; as, for instance, the muscles attached to the joints, the muscles controlling facial expression and the voice. Another set are involuntary, and perform their functions without the control or will of the individual; such as the heart, stomach, lungs, etc. The action of the muscles of the human organism affords a wonderful and beautiful exhibition of mechanical ingenuity and effects, and serves to illustrate one of the fundamental principles of Scientific Physiognomy; viz., that man cannot perform any work outside of himself unless his nature is largely endowed with the very principles which he requires to use in his work. Hence it is that we find in mechanics who excel a good development of the bony and muscular systems; also, in sculptors, orators, and artists. Who ever saw a mechanic with the Vegetative system predominant, with short, fat limbs, large abdomen, small bones and muscles, puffy cheeks, round head, and slow, waddling gait? Such men are incapable, even, of understanding mechanical principles, and wholly incompetent to perform mechanical labor. They are not built upon mechanical principles themselves; therefore unsuited to carry out those principles.

Dr. J. Simms, the eminent Physiognomist, tells us that "a curious law operating in connection with the human faculties is that it is not within the power of any individual to do or perform anything which does not already exist and reside within his own organism. A man need not attempt to become a carpenter or architect, or to build a house, if he is not himself constructed on mechanical principles. If he has not a square form, and is not provided with large bones, he will be quite unable to distinguish himself in dealing with square objects, or things with angles and straight lines." My own observations fully corroborate all that Dr. Simms says on this subject, which he explains in the most simple and concise manner in his celebrated work entitled "Nature's Revelations of Character."

The following description of the powers of the various muscles in the human organism will give the reader some ideas of the mechanical powers and principles included in the Muscular system :

"*The human body combines in itself almost all the principles of natural forces;* viz., the different *lever powers* in the action of the muscles upon the bones. One principle of it is well illustrated in the action of the biceps muscle in flexing the arm; so, also, in the flexors generally—namely, that in which the force is applied between the weight and the fulcrum. Second, the action of the triceps muscles on the ulna, in extending the forearm, is an instance of a lever power where the fulcrum is between the force and the weight. Third, the example of a lever applied to a weight between the fulcrum and the force may be seen in the action of the abductus magnus muscle of the thigh, in abducting the femur. The different joints are well illustrated in the ball-and-socket joint in the hip and shoulder, the hinge joint in the elbow, ankle, and knee. We have also joints with lateral motions, as well as with flexion and extension, in the wrist; a joint with a gliding motion, as in the temporo-maxillary and sterno-clavicular articulations. Then we have the mixed joints, as in the articulation of the sacrum to the iliac bones, the vertebræ, and the immovable joints, such as the sutures, etc. We have also the different principles of valves, in the heart and veins, and the pylorus between the stomach and the duodenum, and the representation of a pulley in the action of the superior oblique muscle in rotating the eye."*

From the foregoing explanation of the action of the muscles within the body of man, it is quite easy to comprehend that, when these principles are largely represented in an individual, he will be better able to understand and put in practice the same principles in objects external to himself than one in whom these mechanical powers are deficient.

From the preceding exhibit of the varied powers of muscular action, it will be seen that this system belongs to the Architectural division of the organism, and, in combination with the Osseous or Bony system (which will next be treated

* Z. Hebert, M. D.

of), constitutes the building powers and capacities of man. Individuals in whom these two systems are well defined are constructive, often artistic, religious, emotional, and amative; with a good quality of brain added, excel in literature as novelists, dramatic writers, and in sensational sermons. Many highly emotionally religious persons are found to be endowed with a fine quality of muscle. It does not necessarily follow that they are moral also. Emotion in excess is opposed to morality. At the same time, muscle assists *faith*, *ardor*, and *zeal*. In the races in which is found the most Credenciveness—that is to say, faith or belief—there will be found the predominance of the Muscular over the Bony system; as, for example, in the Jews, Turks, Persians, Arabians, and Hindoos. And what is true of the races applies with equal force to individuals.

A complete revolution in the science of Human Nature must ensue before we can comprehend the motives and character of man. Herbert Spencer, in his "Essay on Education," remarks that "without acquaintance with the general truths of biology and psychology, rational interpretation of social phenomena is impossible." And he also says, "The actions of individuals depend upon the laws of their natures, and their actions cannot be understood until these laws are understood." Now, the theories of the ancient metaphysicians and theologians were not founded on an intimate knowledge of either physiology or the laws of mind, as revealed by investigation of either bony, muscular, or brain and nerve system. They were, most of them, simply speculative theories, which had no basis in fact and no foundation in reality. They were like the loves of the poets—creatures of the imagination merely. If we desire to advance in exact knowledge of *real human nature*, we must cast out the ancient dogmas, which, venerable as they may be, are untrustworthy, and interrogate nature face to face. First, learn the facts; then manufacture your theory in accordance therewith. Formerly the method was to construct an abstruse theory, couched in incomprehensible terms, and let the facts shift for themselves.

Then the necessity for *faith* and *belief* arose, and was at one
time considered the crowning virtue of human character.
Now childish credulity is looked upon in adults with a pitying
eye, and we feel both sorrow and contempt for him who is
too weak or too cowardly to grapple with the truths of Nature
lest they overthrow some time-honored error which he is
cherishing.

> "For Faith, fanatic Faith, once wedded fast
> To some dear falsehood, hugs it to the last."

When we wish to understand the emotions which play so
important a part in the drama of life, we must look to the
physiological and anatomical conditions of individuals and
races, for it is to the Muscular system mainly that we are indebted for the power to manifest *emotion*, *will*, and *expression*.
The great number and variety of the muscles of the face
alone, where expression is most manifest, will vouch for the
truth of this statement.   The eye alone expresses more feeling, emotion, will, mental energy, and capacity, than all the
other muscles combined.   I do not say that it performs more
labor, but that it *expresses* more of the physical and mental
characteristics of the individual than any other portion of
the muscular system; and the reason why it does this is explained by the fact that the eye is a *mass of muscles*.   Added
to this power is the fact, also, that the optic nerve finds its
centre and seat here.   The eye not only brings the world
into the mind of the individual, but also shows *to* the world
the *mind of the man* as he stands before our gaze.   The muscles of the eye and the optic nerve combined bring to us the
bulk of the knowledge we acquire.   It is true, we can feel
heat and cold, changes in the temperature of the air; we can
taste, smell, hear, and touch without the use of our visual organs; but the world of Form—of Architecture—is unknown
without this sense.   The sense of color and the knowledge
of form bring to us our most useful and practical acquire-
ments, and to the use of the muscular system are we indebted
for much that is practical, useful, and necessary.   Hence,

the importance of endowing our offspring with a good share
of this system. It is a fine inheritance to leave them. It
can be improved by food and exercise, rightly applied, and
those who make gymnastic exercises a daily duty and pleasure are laying up a store of goodness, which, whether they
will or not, will be transmitted to future generations, and
thus "do our deeds follow us."

When we examine the nature of muscle, we find, although
it is powerful in *expressing emotions*, it is wanting in feeling,
in sensitiveness. Hence, we find that muscular people, although able to express emotion, have very little of that keenness of sensation which those have who enjoy a fine nervous
organization. Emotion is not sensation, and thus it is that
those who seem to feel the most in reality feel the least. If
one could cut a muscle without striking a nerve, there would
be little, if any, feeling experienced. It is only by analyzing
the constituents and nature of the several systems in the
body that we are able to give to each its own appropriate
share of work. This method enables us to relieve the brain
of a large share of the labor which former theories of the mind
have ascribed to it. Hitherto, it has been a poor overworked
organ. If the brain is capable of all the labor which has
been assigned to it by metaphysics, of what use, I ask, are
the several ganglia, plexuses, the muscles, and the visceral
organs? We must either divide the labor equitably or declare the utter inability of the last mentioned in assisting
mental manifestations.

THE OSSEOUS OR BONY SYSTEM.

The Osseous or Bony system is known by height, large
joints and bones, high cheek-bones, and predominance of
the lower part of the forehead—projecting over and beyond
the eyes—prominent chin, large hands and feet, and moder-
ate sized brain. The bony structure is the foundation and
framework of the human organism, upon which is built the
entire man; and to the predominance of the bony structure

ANDREW JACKSON.

man owes his character for integrity, and stability, and physical and mental soundness; the very constituents of bone —lime, phosphates, magnesia, soda, etc. —give stability, integrity, decision and firmness to the organization in which they abound most largely; hence, the Bony system is the one in which, *from the very nature of its components*, we naturally look for the most stability and trustworthiness. It is also built upon the *straight* principle, "the bones forming right angles to each other, causing the character to accord with its *upright* and *downright* architectural formation." (Simms.)

Long, lean, bony people are noted for their usefulness, unselfishness, integrity, and generally for mechanical ability. Bony people, with a fair proportion of the Muscular system in combination, make the best mechanics in the world; length facilitates activity, while muscle, combined with a large bony structure, gives the power essential to mechanical construction.

This system is included in the Architectural division of the face and body, and has for its assistance the muscular powers. These two systems combine and include most of the principles of natural forces, as has been shown. Now, persons in whom this combination is largely developed will not only have the power to become good mechanics and artists, but will be able to build up and perpetuate a fine race of children, if proper attention be given to combinations with suitable conformations, added to righteous regard for hygienic and sanitary laws. It will be perceived, from this analysis and illustration of the Bony system, that the human organism is dependent upon an excess of bone development for all those attributes which go to form stability, integrity, as well as architectural and mechanical ability. These principles lie at the very foundation of physiology, anatomy, human greatness, moral goodness, government, and society; and in every age, country, or community noted for its justice, probity, and true civilization, there will be found, upon examination, a majority of its people built upon this conformation and possessed of mechanical powers.

In selecting trades for young people, due attention should be paid to this principle of Nature. A neglect of its application would result in failure; and one reason why we sometimes find poor mechanics is that they have mistaken their vocation, and chosen a pursuit to which their conformation was unsuited.

The signs of the bony form predominant are found *all over* the individual—in the large joints of the hands, fingers, wrists, arms, and legs. The projection downward of the lower jaw, forehead projecting over the eyes, the high, bony nose, are all evidences of a conscientious and morally inclined character; indeed, the Bony system may be depended upon for moral conduct. The large development of bone shows that the fluid circulation has done its work in a thorough manner, and has conveyed all the materials needed in bone-making to their several destinations, in just the right proportions, thus giving soundness to the whole framework. Size and Form, Physical Order, and Calculation are some of the prominent faculties in this system, as well as Veneration and Executiveness. Conscientiousness is seen all over the individual in whom the Bony system predominates over all the other systems. The list of faculties in the Architectural division are as follows: Secretiveness, Force, Resistance, Hope, Cautiousness, Imitation, Analysis, Ideality, Sublimity, Human Nature, Constructiveness, Acquisitiveness, Veneration, Self-will, Executiveness, Credenciveness, Observation, Memory of Events, Form, Size, Weight, Color, Order, Calculation, Locality, Music, Language. Some of these faculties are derived from the functional action of the heart and lungs; others from the power of the muscles and bones; Hope is related to the liver; Veneration, to the stomach. Every faculty depends upon some organ or system of functions for the power to exhibit its peculiar mode of activity.

This page concludes the description of the Architectural division of the face and faculties. The term "architectural" is used in its broadest and most comprehensive sense. Whatever exists has a form—is *built*; not a particle of any sort whatsoever is found without form and without combination with some acid, gas, ether, or solid substance; thus it is architectural. In works of art, the same principle applies; in dramatic composition, in works of fiction, and in the sermons of the preacher, the same *mechanical constructive element prevails*. Constructiveness is the basis upon which all depend, and wherever we look—whether into the world of material organization, the universe of forms, or into the world of thought—the same formative or architectural law governs and controls.

THE BRAIN OR NERVOUS SYSTEM.

One of the best evidences of a developed race is found in its manifestation of high mathematical powers. The unperfected races of the world, among whom I may mention the South Sea Islanders, many African tribes, and the Esquimaux, have so little ability in this direction as not to be able to calculate or comprehend anything beyond the number of their fingers and toes. I think that man's superiority over the brute creation is more marked in this respect than in the matter of simple reason, which attribute many deny to the animal kingdom, although the power of reasoning, to a large extent, is proved by modern naturalists to hold a place in the mentality of the higher races of animals.

The faculties of reason —Causality and Comparison—endow man with the gift of abstract mathematical ratiocination; in this respect he is, perhaps, more distinguished from the brutes than in any other manner, with the exception of the faculty of Speech, although this is possessed by the parrot; but in this instance, speech proceeds from a suitable formation of the vocal organs and a good development of the faculty of Imitation, and

HERBERT SPENCER.

is not accompanied with a corresponding degree of sense or sensibility. This might serve as a lesson to those persons who ascribe to the human race "divinity," and to the lower animals none. When we find the lower animals endowed with a fine degree of reason, as in the horse, dog, and elephant; and some mathematical ability or sense, as in these same creatures and in "learned pigs," who are taught to count and reckon; when we find human speech in the parrot, I think the self-love and vainglory of Man may as well give way, and allow to these, our "blood relations," as Haeckel terms them, a fair share of divinity. We ought to be thankful to the Creator, who, in His wisdom, has chosen these humble instruments to serve as a means of teaching us whence we sprung and from whom we originated.

The Brain system is the highest and last in the progressive development of the human race. Its physiognomical manifestation is shown by breadth, fullness, and height of the forehead. The annexed portrait of Herbert Spencer gives a representation of its shape and position in the physiognomy.

You will observe that the brain and nerves form one system and belong to the Mathematical division, and its work is performed by these *two* forces, which are really *one* force, but usually considered two on account of the difference in locality, the substance which composes the nervous system being identical with the brain matter. Those in whom this system predominates are noted for their intensity of thought and feeling, refinement, mental acumen, memory, desire for learning, pure and virtuous desires, great physical activity, sharp features, rather small nose, thin nostrils and lips, face of a pyriform shape, small bones and muscles, slim build, too apt to overdo and exhaust the vital forces. Many of our most brilliant thinkers have possessed this system. Many, not having a due proportion of the other systems, have died young. This conformation large, in combination with the Bony and Vegetative systems harmoniously blended, has produced many powerful intellects. Samuel Johnson, Arkwright, Gibbon, Dumas, Buckle, Hume, Benjamin Franklin, William Penn, Handel, and many others, were thus endowed and were noted for their thought and industry. Individuals in whom the Brain system has the ascendancy are liable to dyspepsia, nervous exhaustion, and irritability, apt to be fussy and exacting in small things, and desire to exhibit taste in all their surroundings. The faculties originating from this function are Causality, Comparison, Time, and Intuition. The reader will doubtless say, upon observing the relatively small noses of persons in whom the Brain system preponderates, that the law laid down in the chapter on sub-basic principles is inconsistent and cannot be harmonized with the facts observed. Let us examine the evidence in this case. The law states that "the size of the nose, controlled by *quality*, is the measure of power." Now, if *quality* were not must potent in deciding power, the noses of the most gifted persons would have to be enormous in order to show the *power* of the traits whose signs are located in the nose—particularly about the tip of the nose, where so many signs are placed. As the human mind and body rise from the Vegetative system up and through the several higher systems of functions and faculties, the size of the nose decreases (relatively) with the increase of quality; yet, where a large nose coexists with fine quality, a first-class intellect in some direction is indicated. The peculiar characteristics must be decided by the *shape* or *form* of this member. The fact is, very gifted persons make up in quality what they lack in quantity or size, and thus the nose is able to indicate many beautiful traits in a small space.

The intelligence of low animal organisms is exhibited by the nervous system which they possess, be it simple or complex. As these organisms rise higher in the scale of progression, a simple brain is formed. This, in course of evolution, rises in the grade of intelligence and assumes many diverse forms, until a more complex brain is evolved, such as we find in civilized man. Until the brain appears, all knowledge of the outer world comes to the lower organisms by the sense of touch mainly. This progress from the simple to the complex epitomizes the different ascending grades of intelligence in man. The lowest form is found with the purely Vegetative system. Rising a little, the Thoracic lends its aid, by the simple fact of a greater degree of oxygenation of the blood, and by increased power of the heart portal and circulatory systems. Next in the grade of increased usefulness we find the Muscular system bringing a rich endowment of functions and faculties; also, of greater procreative power, for wherever we find a large development of the muscular and fibroid systems, the ability for *procreation* and *creative* effort is augmented. This is the law right along the line of progressive growth: *The more physical functions, the more mental faculties.* Following this natural order, we next find the bones, formed *chiefly* by the action of the sun, although food and water contribute their quota. This Osseous development gives mechanical and inventive powers, either in man or animal. Last of all, the more perfected Brain system makes its appearance, together with a more complex arrangement of the nervous system. This produces greater activity

5

of thought and reason, and consequently of mathematical power. How any one can study the human physiognomy scientifically, and not find therein the proof of the evolution of the world—of the animal and human races—passes my comprehension. It is a "revelation," a "gospel," and a "religion" all combined,

The three divisions of the face represent, also, the three ages of man's existence—infancy, middle life, and old age. They also illustrate, in the manner of their unfolding, the methods of Nature in evolving a *race*. The features in infancy are like the lower types of men, flat and almost expressionless; in middle life, the central division comes into full expression and activity; later, after the mind has been strengthened and stored with fact, fancy, and practicality, pure reason asserts its dominance by enlargement of the cerebrum, until old age comes on apace and dissolution ensues, typifying the decay and death alike of man, of races, of nations, and of the world.

CHAPTER IV.

LOCALITIES AND DESCRIPTION OF SIGNS IN THE FACE.

"Racial history, ancestral characteristics, personal aspirations, passions, and exploits become part of each living soul, and are portrayed distinctly on the outward form, and more especially on the most discernible portion—the countenance."

—Dr. J. Simms.

The form of the human body is only one of the many indications of human character. The voice, the walk, gesture, attitude, handwriting, and hand shaking are all indices and exponents of traits, as well as of physical and mental conditions. A skillful and observant person can tell much by the hand or foot alone; by the eye, very many things are indicated; the nose reveals much of the mind and interior of the body; in short, each feature has in it many meanings. In the pages which immediately follow this, the way to discern and locate the signs of the various faculties will be explained.

The use, primarily, of all the functions and faculties is for the preservation, protection, and perpetuation of the species. Other faculties and powers have aggregated by use and attempts in higher directions. Practice increases capacity. There is no doubt that the human mind is gradually acquiring more faculties by striving after higher knowledge. These, undoubtedly, will be evolved in the regular order of progress, from the lower to the higher. The present age is expanding and strengthening the higher powers of the mind; reason is more general among the civilized races than in any previous era. As a consequence, superstition is giving way to positive scientific truth and demonstration, and theories unsustained by reason and fact are impeached and rejected.

As the powers of the mind expand, we become cognizant of facts in Nature which lower developments failed to perceive and could not penetrate. We are gradually, but slowly, becoming acquainted with the world we live in, and things which have seemed to be the work of supernatural powers are now so well understood as to come within the comprehension of school children, and can no longer be used to pander to the ambitions, vices, or designs of wicked kings, crafty priests, or unscrupulous politicians. Among the most important discoveries, I may mention the science of Physiognomy, which is destined to play an important part in the civilization of the world, by unveiling what has been so long a mystery to man; viz., Man himself.

Physiognomical sensation, as Lavater designated the innate and intuitive conception of character, is common both to men and animals. A dog will show by his actions that he understands character, and will be instinctively attracted to those who love his kind. Babes, who are yet in the stage of animal instinct, will attach themselves at sight to those who are fond of children. Men, in looking at the faces of others, will be drawn in confidence or repelled by something in the countenance which they cannot define or locate exactly.

They say of one, "He is a good, square man;" or, "He is a sneak and coward—I can tell it by his face;" and yet, if you ask them to point out the precise places where they discover these traits, they cannot tell you where they are to be found.

The possession of this physiognomical instinct is general, and shows not only that the face is understood to be for some other purpose than to place the eyes, nose, and mouth conveniently, but instinct and intuition as well point to it as the natural record of the body and mind—of the real Man himself. The nerves of sensation ramify upon the face and front of the organism, while the motory nerves are at the back of the brain. This disposition of the nerve forces would cause the face not only to exhibit more of the character than any other portion of the body, but would prove the fact that the greater the development of the features of the face, the greater its power for receiving sensation; thus exhibiting more gifted characters than where the features are small and undeveloped. All human nature attests this fact, and shows that the more varied are the features—the more depressions and elevations there are in the face—the greater the variety of character is exhibited. A smooth, small featured, un-wrinkled face always discloses a small, unemotional, unthink-ing, and selfish character, of very small capacities. A man's real character is spread all over him. His voice and walk agree with the shape of his body, and reveal his mentality to a degree; but the face sums up the whole Man.

As I have before shown that certain powers are derived from the predominance of certain conformations of the or-ganism, and are always found accompanying them, it is log-ical to infer that *determinate portions of the body sustain and are related to certain faculties of the mind.* Upon investigation, it will be proved that the face is the exact register of all mental faculties and bodily functions and conditions. A keen analysis and comparison of the development of the organs of the body with the action of the faculties, emotions, and sentiments, will show that the organs of the viscera—the kidneys, the reproductive system, the liver, the intestines, the heart and lungs—as well as the bones and muscles, sustain and are directly related to certain mental faculties. All mental faculties have their physical bases, from which the mind is able to produce thought, emotion, or will. This interaction of the mental and physical powers will be explained as we proceed. The locality of signs in the face will here be given. The *rationale* of the order of their arrangement will be made apparent as the reader progresses.

CONSCIENTIOUSNESS.

SIR ISAAC NEWTON. (Conscientiousness.)

Conscientiousness is largely represented in the face of this eminent astronomer and scientist, and assisted him in comprehending those truths and laws of Nature which he had discovered.

Rectitude of character is known in the face by that width of chin which is produced by the development or width of the inferior maxillary, or lower jawbone, and by a general squareness of all the bones and straightness of the face; also,

by the manner in which the eyes are placed in their orbits.
Eyes which turn far downward at the upper and outer corners are not truthful eyes. Those which are almond-shaped
or cat-like, and turn upward at the outer corners, are crafty
and deceitful, as seen often in the Mongolian race. Indeed,
any eye off the straight line varies in truthful signification,
according to the amount of its deflection from a straight line,
running all the way from amiability through the various
degrees of plausibility, duplicity, deception, secretiveness,
craft, cunning, lying, and cruelty; all of which are shown by
the shape of the eye which deviates from a straight line,
either above or below the line. Where the outer corners of
the eye turn upward, the indications are like those of the
same shape in the lion, tiger, fox; and like traits will be
exhibited—cruelty and craft, deceit and cunning. But where
the outer corner turns downward slightly, agreeability of
speech is always found; indeed, persons possessed of this
sign would rather tell a pleasant untruth than be the bearers
of harsh or unpleasant tidings. Still farther turned downward, they are plausible and persuasive, and make good
salesmen and politicians. Benjamin F. Butler's eyes are
more marked in this respect than the eyes of any celebrated
man that has come under my observation.

As Conscientiousness gives moral courage, it is the base
of many heroic acts, and will often lead even delicate women
to deeds of daring for principle's sake, and to protect the
helpless. Joan of Arc and Charlotte Corday are examples
of the former; and Mrs. Patten, the wife of Captain Patten,
the lady who navigated the ship "Neptune's Car" from
Bombay to Boston, while her husband lay delirious in his
berth, is an instance of the latter. This lady nursed her
husband, navigated his ship, and suppressed a mutiny with
the assistance of part of the crew, and carried the ship safely
into Boston Harbor, a few years since. This woman was a
small, delicate person, and gave birth to a son a few months
after this occurrence. Her portrait shows that Conscientiousness was one of her leading traits. Such women are
the mothers of heroes.

This faculty is large in the faces of North American Indians, who kill and slay from a sense of justice (knowing themselves wronged by a Christian government), and not, as many think, from a love of slaughter. Under an equitable system of dealing, they have ever proved true; notably in their transactions with the Quakers under William Penn, who always made just and satisfactory terms with the Indians, and then kept his agreements rigidly.

This faculty in excess leads to severity and exaction in moral conduct and life. The pioneers in all departments of advanced thought, in governmental and moral reforms, have possessed this faculty largely. The faces of Franklin, Jefferson, Jackson, Washington, Paine, John Bright, Cobden, Wm. Cobbett, William Lloyd Garrison, Abby Kelly Foster, Frances Wright, Lucretia Mott, and all who have dared to demand the abolition of unjust laws, and who have contended for the establishment of new forms of government based on human rights, evidence by their physiognomies that Conscientiousness fills a large part of their natures.

Conscientiousness is found most largely developed where the Bony system predominates; and, as liquids do not affinitize as well with this system as with the Vegetative and the Muscular, there is, consequently, less drunkenness among persons of the Bony structure predominant. Many leaders of the temperance movement will be found to possess the Bony system predominant. Those who have been great drunkards and have reformed, like Gough and Murphy, the leader of the "Murphy movement," are men of muscular build, and are held to their pledges through their religious associations, and do not depend upon pure Conscientiousness, which is found most active with persons of the Bony system.

Conscientiousness is in the domestic group, and does not belong to the religious group of faculties. *It antedates them in the evolution of organs, functions, and faculties*, and is of far more importance in the human organism, being primarily for the protection of the purity of the entire body. Conscientiousness is related to the kidney system, which both secretes and excretes the fluid waste and impurities of the entire body. As seventy-five per cent. of the organism is water, the physiological importance of the organ must be apparent at first glance; its moral importance follows as a matter of logical sequence. The relation of the various organs of the body to the moral and social faculties is explained fully in the chapter on "The Rationale of Physical Functions and their Signs in the Face."

FIRMNESS.

ANNA DICKINSON. (Firmness.)

The portrait of Anna Dickinson exhibits this faculty in a marked manner. She is noted for her perseverance under great difficulties. She is also a celebrated orator and writer. In her youth she was a poor working-girl, but with the aid of this faculty has achieved eminence and distinction.

This faculty, shown by length of the lower jaw, evinces great tenacity of purpose, stability, perseverance, and decision. A retreating chin shows a lack of all these qualities. Firmness is an attribute only of developed races and individuals; man is the only being endowed with a chin. All of the undeveloped races of men have narrow, retreating chins and depressed noses. (Observe the faces of the New Zealanders.) The length of the chin is one of the facial indications of the Bony structure, and, in combination with Conscientiousness (width of chin), is the base of the heroic. Indeed, this faculty (Conscientiousness) is the primal cause of moral action; true heroism could spring from no other motive. It also gives the power for fidelity to principle, truth, and justice. The chin is the seat of heroic character, which depends upon the firm and substantial nature of bone for its sustenance.

A certain writer on Physiognomy—Redfield—has given the chin as the locality of the signs of Love. He certainly could not have considered the nature of bone in this connection. We do not love with our bones. Love signs are found predominating in those who are the most emotional and impressible. Muscle and fat are more easily acted upon than bone, and the physiognomical signs of Love will be found in the muscles of the face, just as Love is found more largely developed in muscular organisms.

Bone shows more of integrity; muscle, more of the affectional nature. Length of chin indicates perseverance and calm, firm, persistent action, rather than what is called will-power, which exhibits itself in sudden bursts of violent temper. This distinction must be thoroughly understood. The nature of bone, like that of rock, offers a steady resistance and pressure, and large Firmness is the result of a large development of the Bony system. Muscle has a reactive quality, and will-power is based on and operated by muscular action purely.

The physiognomy of Miss Anna Dickinson admirably illustrates the faculty of Firmness. She has pursued the career which she marked out with a firm and unyielding determination. It is not singular that success should crown her efforts; she stands to-day pre-eminent among the women of the age, in mentality, in moral and intellectual power, and in works of humanity. The scientific physiognomist knows well how much she is indebted to a good, honest Bony system for her power. Had she possessed an organism composed of soft, fat, flabby material, she would never have been heard of outside of the kitchen and dining-room, the favorite resorts of this description of persons.

DIGESTION, OR ALIMENTIVENESS.

ALEXANDRE DUMAS.   (ALIMENTIVENESS.)

The facial expression of Alexandre Dumas, the novelist, illustrates the function of Digestion, or Alimentiveness. To an active brain, he added large assimilative capacity. This enabled him to perform very arduous mental labors. His writings are extensive; his capacity for both solid and liquid foods was enormous.

The mouth being the entrance to that chemical laboratory, the stomach, large size would disclose great appetite and power for digestion. Fullness of the lower part of the cheeks is another sign of good digestive power, for if the food assimilate with the juices of the stomach, liver, and pancreas, the lower part of the cheek will indicate this condition. Large development of the parotid or salivary glands, just in front of the ears, is another proof of assimilative power. When this gland is well developed, a soft, cushion-like protuberance will be observed directly in front of and below the opening of the ear. It is usually large in outdoor laborers, seamen, farmers, and all who eat heartily and digest well.

BENEVOLENCE—GENEROSITY—SYMPATHY.

WILBERFORCE. (Benevolence.)

The picture of Wilberforce, the great philanthropist, illustrates this faculty in every particular. He was instrumental in abolishing the slave trade of the English colonies, and imperiled his position in a courageous protest against the unchristian mode of life practiced by the middle and upper classes of England. Died 1833; was buried in Westminster Abbey.

The full, rolling under lip is a sure sign of these faculties; it depends on the faculties in combination which of these forms will exhibit itself. I have known one with this indication large, but with small Friendship, so destitute of sympathy as to leave the sick uncared for, yet loved to give money and presents, apparently for the sake of giving, just to please himself. In others, especially in combination with Friendship, it takes the form of active sympathy. In conjunction with Conscientiousness and good reasoning faculties, it shows true benevolence or philanthropy. It depends entirely upon the faculties in combination with it as to the particular way in which it will outwork. The philanthropic phase of this faculty is seen in the portraits of Elizabeth Fry, John Howard, George Peabody, and Peter Cooper. In all of these persons, Reason and Conscientiousness, combined with a broad understanding, induced a desire to benefit large numbers of their fellow-beings in one broad plan or scheme of benevolence. The lower lip of stingy, close, miserly people is usually very thin and dry. All the noted misers whose portraits I have observed have this peculiarity. Abraham Lincoln's portrait illustrates the faculty of Benevolence thoroughly. His life-work is the proof of his facial record. The action of this faculty and function is derived from the glandular system.

AMATIVENESS.

This faculty is directly related to the reproductive system. Its principal facial physical sign is found in the lip. The middle of the upper lip, if very thick and red, is indicative of large passional and reproductive powers. Another sign of Love is found in the eyes; the larger and fuller the eye, the greater the devolpment of the love-nature. I have found that the fullness and size of the eye is more significant of the sentiment of Love; the development and color of the lip shows its physical power and activity merely. Small eyes evince much less amative power than large ones, and are often accompanied by a thin, white upper lip. The shape of the eye is potential in disclosing the kind of Love the indi-

N. GOODWIN, COMEDIAN. (AMATIVENESS.)

Mr. Goodwin's countenance is a fine exposition of the faculty of Amativeness. He is very original and creative in his efforts as an actor. The Glandular and Muscular systems, from which Amativeness is evolved, assist in creative art. He has a fine endowment of these systems, as have all artistic persons who excel.

vidual is endowed with. The round eye, like that of the dove, is an unfailing sign of large mating ability, of faithful conjugal attachment, love for one only. In some instances, it is so marked as to prevent the individual from ever forming but one attachment for the opposite sex in a life-time. The thick upper lip and full eye are also indices of muscular power. Eyes the commissures (or openings) of which are almond-shaped prove promiscuous love. For these facts in regard to the form and size of the eye I am indebted to the renowned Physiognomist, Dr. J. Simms. The eye is significant of many other characteristics which will be treated of in their proper places.

LOVE OF CHILDREN AND ANIMALS.

Love of offspring and of the young of the animal kingdom are traits which are seldom seen separated. The most prom-

FRIEDERICH FROEBEL. (LOVE OF YOUNG.)

The founder of the Kindergarten system of education possessed a wonderfully intuitional comprehension of child-nature. His system is in accord with the natural development of the human mind, commencing with the mechanical and artistic faculties and followed by the use and development of the reasoning powers. His love of children and mirthfulness, his ingenuity in contriving plays and methods, and, above all, his insight into child-nature, rank him among the first educators of the world.

inent sign is found on either side of Amativeness, causing the outer corners of the upper lip to droop and form a slight scallop. Where this faculty is large we observe the formation of a "Cupid's bow" in the contour of the upper lip. This sign is very marked in dogs, horses, and cows. It is so large in these creatures that the outer sides of the upper lip project downward over the under lip. Their love of young is intense. In horses and dogs this feeling extends to the young of the human species. Many dogs, which are ferocious to all adults, have been known to caress and treat with kindness young children. The development of the mammary glands is an unfailing sign of this sentiment.

MIRTHFULNESS.

ELIZA COOK. (MIRTHFULNESS.)

This happy, mirthful, hopeful countenance of the author of "The Old Arm Chair," and many other lovely poems, portrays in every lineament the dominance of the fun-loving and fun-making propensity. Love of young and hopefulness are also very decided traits.

Mirthfulness always accompanies large Love of Children, and its sign adjoins the latter. The harmony of Nature is exemplified in a wonderful manner in the grouping and associating of faculties. Mirthfulness is essential to the love, care, and companionship of children, in order to attract their love and company and to amuse and entertain them, and in the face its principal local sign adjoins that of Love of Young. The sign for this faculty lies just outward from the corners of the mouth, next to Love of Children, and between that faculty and Approbativeness. It is shown by several small vertical wrinkles in the muscles at this point, and also by the corners of the mouth turning upward, and by a merry, pleasant expression of the eye, as well as by several straight wrinkles running outward from the corners of the eye. This last sign is seen only in adults. A gloomy and sad disposition is never found with Love of Children, and when we observe the sign for Mirthfulness in a face we naturally look for Love of Children and Animals. When Nature creates a faculty, such as Love of Children, Love of Music or Art, the faculties needed for its expression in some form or other are provided and work in harmony with it. Under this law of Nature it is safe to predicate, from seeing certain signs in the face, that certain other faculties are present. The origin of Mirthfulness is glandular.

APPROBATIVENESS.

CARL LINNÆUS. (Approbativeness.)

This eminent Swedish naturalist, noted for his contributions to Botany, died in 1778. He was very susceptible to praise and flattery, and very approbative of the efforts of others. His physiognomy is quite marked in this direction.

This is the faculty which gives rise to love of commendation and praise, and which also makes one sensitive to the speech and opinion of others. It causes individuals to be ambitious and to desire to excel, that they may be praised and well thought of. It is distinguished from the faculty of Self-esteem in that it seeks the good opinion of others, and cares more for the applause of the world than for the approval of self. Its action is often mistaken for that of Self-esteem by superficial readers of character. Approbativeness is found larger in actors, artists, and singers than in other classes. It is essential to the success of these people; the approval of their audience is the spur and incentive to still greater efforts. It is also large in those who are fond of fashionable life, who love display and make great exertions to "keep up appearances." This trait leads public men to desire popularity, praise, and attention; and, in excess, makes "shoddy aristocrats" of those who ought to be proud of being American citizens, who should be content to be classed among the "plain people," as the good Abraham Lincoln named the laboring masses. Approbativeness is the incentive to many great and noble deeds. It inspires the orator, artist, and actor with the desire to win fame by the excellence of their achievements. It is a very useful faculty where it is possessed in a balanced degree. Unbalanced, it produces excess of vanity, foolish love of display, and an undue dependence on the opinions of others, thus taking away all true dignity and independence of thought and action. The location of the sign for Approbativeness adjoins Mirthfulness in a straight line outward from it, and often produces dimples at this point. It is shown in others by two or three deep vertical wrinkles at the same place. Where this faculty is uncommonly large these wrinkles are permanent; in others they are observed while the face is animated, as in smiling or in conversation. Approbativeness originates in the glandular system.

G

FRIENDSHIP.

ELIZABETH FRY. (FRIENDSHIP.)

The physiognomy of Elizabeth Fry, the prisoner's friend, shows in every feature the friendliness, philanthropy, and nobility which characterized her life. Her endeavors to mitigate the horrors of prison life, and to lessen the sufferings of the insane, are historical. Her idea of woman's sphere was bounded by the sphere of human suffering. No one can gaze upon this face without venerating it.

Friendship is the faculty which gives love of associating with friends, society, and congenial spirits. The physiognomical signs of Friendship are known by a fullness of the upper part of the cheek, over and below the malar bone. The cheeks, both in the upper and lower parts, seem to be nourished by the action of the bowel system. The *rationale* of friendship is that it primarily seeks the bodily comfort of the objects of its affection; it sets before them food and drink and all creature comforts. In its sentimental aspect it shows more in love of association, in presents, and in delicate attentions. Its excess often produces a silly, gushing demeanor.

SELF-ESTEEM.

CHARLES BRADLAUGH. (Self-esteem.)

Self-esteem very large. Charles Bradlaugh, Member of Parliament, author, editor, and orator, is noted for his determined stand in advocating the right to freedom of conscience. His firm belief in his own powers assists materially his reformatory efforts.

This faculty gives the power for self-appreciation, true pride of character, independence, and self-control. It is one of the noblest faculties, proving often a "tower of strength" to its possessor. It lends dignity and decorum to the demeanor. Those possessing a large degree of Self-esteem are never undignified, even in their amusements, but will be guided by a strong sense of propriety upon all occasions. Self-esteem is more common to northern characters than to the inhabitants of the tropics. It is found large in the countenances of the Scotch, English, and New Englanders. Relative length of the upper lip is the physiognomical sign for this faculty, and this indication is always accompanied by a straight, erect carriage of the body and head. The faces of

Washington, Sir Walter Scott, Admiral Farragut, Herbert
Spencer, and John G. Whittier, all well known to the public,
have the sign of Self-esteem large. Where this faculty is
found in a balanced degree, there is much in the character
to esteem and admire. The more the reader investigates
the human mind scientifically the more he will be convinced
that Nature works intelligently, and never places a faculty,
or sign for a faculty, in an organism without a use and meaning. An excessive development of this faculty leads to ego-
tism and a feeling of importance; makes its possessor very
grave and ludicrously dignified; he sets great weight upon
his opinions and desires to be treated with marked respect
and consideration. The sober faces of some baboons, who
have an extremely long upper lip, are quite amusing in their
gravity, while their actions are anything but dignified.

MODESTY.

SIR WALTER SCOTT. (MODESTY.)

Sir Walter Scott inherited, with his great talent, an exceedingly delicate and sensitive organism. His uncommon modesty is remarked by his biographers.

Modesty is found in those who possess an uncommon sense of purity of thought and action. One of the most prominent signs of Modesty is the channel-like groove running down the centre of the upper lip, from the septum of the nose to the extreme end of the lip, where it terminates in the sign for Amativeness; it is most developed where the depression is well marked. This faculty is found large in all those in whom the Brain and Nerve system predominates, *whether this sign be present or not.* The Brain system always gives a more elevated cast of thought than the other systems. As its position indicates it to be the highest in the organism, it will be found to produce the purest and loftiest sentiments and emotions. It is the system of quality and fineness; hence, fine hair and skin are also signs of purity, modesty, and love of cleanliness and neatness. The faces of Charlotte Bronte, Lucretia Mott, Elizabeth Barrett Browning, Beranger, the French poét, and Lavater, are well marked in this respect.

All of the foregoing faculties and signs of character are included in the Chemical, or underlying division of the organism, and are mainly dependent for their activity on the quantity and quality of the juices secreted and fibres and tissues supplied by the several organs which compose the systems from which their powers are obtained. After the full and explicit exposition which I have made of the nature and action of these functions and faculties, it would seem superfluous to call attention to the fact that, if a moral, upright, healthful, and happy life is desired, all of the physical organs in this division must be used with a conscientious regard for their care and preservation, for on them depends mainly our power for usefulness and morality; and any system of ethics which ignores this as a fundamental principle will save neither mind, soul, nor body, here or hereafter.

In localizing the signs of the faculties, Phrenology was obliged to encroach upon the domain of Physiognomy; and yet, after having accepted and taught the fact that the face indicates character, by localizing the "perceptives" and "reflectives" in the upper part of the face, one of its expounders, Dr. Grimes, says: "After having studied Physiognomy for many years, I confess that little can be known by the face alone;" while another, O. S. Fowler, states his belief in Physiognomy, and hopes that a system of this science will some day be discovered. Mr. Fowler says, further: "The mind and face are inter-related; all the mental operations shine out through the human face divine. Highly emotional persons manifest themselves more emphatically and distinctly by their countenances than words. Peculiar shades of feeling and existing thoughts and desires are expressed and can be read in this 'mirror of the mind' better than words can possibly portray them, and without the possibility of deception in the one read or reading, and without instruction by either. And, since some can be thus read, all can, of course. Indeed, *facial expression is by far the best medium known to man.*"

In giving the locality of the signs of the faculties comprised in the Architectural division of the face, you will find that there are several which phrenologists classify as "perceptives" located, in this scheme of Physiognomy, in the same places which Phrenology has set for them. My theory of their origin and action, however, is entirely different to that set up by the phrenologists. They discerned and located many local signs of character, situated in the upper part of the face, which scientific Physiognomy endorses as to their locality and meanings, but utterly repudiates the idea that they derive their power from the *brain alone.* These signs are indications of the conformation of the Bony structure; as, for example, in the signs for Size and Form. These are illustrated by width of the Bony structure, and not by any "organ" of the cerebral substance. Others located in this region are caused by the Muscular system; as, for example, Language and Locality. The theories of the phrenologists are wrong, while their observations, in some instances, are correct. They make their best tests, however, in the *upper part of the physiognomy*. Theirs is a system of *half truths*.

In accordance with the plan of Nature upon which this system is founded, I shall here present the signs of the following faculties: Force, Secretiveness, Resistance, Cautiousness, Hope, Analysis, Imitation, Ideality, Sublimity, Human Nature, Constructiveness, Acquisitiveness, Veneration, Self-will, Credenciveness, Observation, Form, Size, Weight, Locality, Color, Order, and Calculation. These faculties represent all of the architectural, building, artistic, literary, and religious powers in man. Conscientiousness is not a religious faculty, as many suppose, but a moral one, and is not found in the face in the religious group, but with the Chemical or Domestic division, in the very basis or beginning of the functions of life, and, no doubt, is formed while the germ of life is in a watery or fluid condition, as it is found, in the perfected human organism, to be the representative of the kidneys, or fluid system of the body.

FORCE.

The power to destroy is evidenced all through Nature by the size and form of human organisms. There are many signs for this faculty both in men and animals. Those which are common to both these classes consist in the predominance of the Muscular system, breadth of chest, width of mouth, wide nostrils, and an abundance of coarse hair. The facial signs of this faculty in man are large size of the mouth, heavy lower jaw, wide flat nose, high cheek bones, low forehead, and relative small and very wide head. Force is one of the most essential faculties. Its adaptation is, primarily, to the destruction of beasts for food, to fishing and hunting to obtain the means for sustenance. Without this destructive power humanity could not progress; as, for instance, in the tearing down of mountains, in excavating rocks, and in all the operations essential to the progressive development of the country, all of which involve destruction before the

DEAF BURKE, CHAMPION OF THE ENGLISH PRIZE-RING.　(Force.)

process of building can be commenced.　Its excess leads to severity, useless destruction, revenge, violence, and murder.

SECRETIVENESS.

Secretiveness may be recognized by compressed and thin lips, and small mouth if the *lips* are thin.　There are other signs of this trait: small eyes, shy and sly glances of the eyes, looking downward and out of the corners of the eyes, and long eyelashes, all indicate different degrees of secretive inclinations. Broad, flat nostrils are still other signs of secrecy. This sign is common to negroes and all undeveloped races. As the large eye and large mouth are evidential of volume of language, so the converse of this will show a lack of linguistive desire. All orators have very large mouths. Secretiveness is an important faculty in the human organism. Without a due degree of secrecy, we could not be just to our friends, nor plan and manage our own affairs with interest to ourselves. An excess of Secretiveness causes slyness, cunning, deceit, falsehood, cruelty, and treachery. Many animals possess a large share of this quality; it is essential to the preservation of their lives, and assists them in gaining a subsistence.

A well balanced mind, possessed of Conscientiousness and good reasoning faculties, will find little use for this faculty.

ROBESPIERRE. (SECRETIVENESS.)

Secretiveness is well illustrated in the features of the tyrant Robespierre, who deluged France with the blood of his victims. He was cruel, crafty, and secretive. Guillotined 1794. Vindictiveness and a lack of Benevolence are noticeable in his countenance. His life corroborates his physiognomy.

Its operation is explained elsewhere. When largely exhibited in Man, it denotes a great deficiency of either mechanical, moral, or intellectual capacity, and is the compensation
Nature gives for these defects. All animals who exhibit
great Secretiveness are deficient in power of some sort, and
Secretiveness is, in their case, the faculty which contributes
to their means of subsistence. This faculty is large in
snakes, rats, opossums, foxes, hares, and all other animals
who are weak or timid. Among men of power and capacity,
it is never found in a marked degree; they do not need it.
A character like George Washington's, for example, could
maintain itself in every position of life without resort to this
trait, because he was possessed of mental and moral resources to meet every emergency, and hence had very little
use for cunning, craft, or deception.

RESISTANCE.

See portrait of Sir Charles Napier at Executiveness.

This faculty, like all the others found in the human organism, is difficult to designate completely by any single word
in our language. Sometimes it shows itself by a combative
disposition; at others, by resisting assaults, by courage, in-
trepidity, resolution, and thoroughness. It gives force to
mental energies and physical prowess; it assists the preacher,
moral reformer, and temperance lecturer to enforce their
ideas in a vehement manner. It also is the power which,
when perverted, gives the pugnacious and quarrelsome their
force and combative disposition. It is indispensable to
every character; it gives presence of mind and coolness of
judgment in danger. There is scarcely a day of our lives in
which we have not need to invoke its power in some form or
other. Life is one long round of resistances. We resist
aggressive infringements of our natural and acquired rights;
we resists the elements, and erect barriers to protect ourselves against the assaults of wild beasts; we resist the en-

croachments of disease, by applying the remedies with which
Nature's great laboratory has supplied us; in short, Resistance gives us the power to *live* under all circumstances.
Without it, we could neither gain a livelihood nor retain our
health. Its excess leads to aggression, bullying, fighting, and
war. Some observers give, as one sign of Combativeness,
Resistance, or Courage, the ears standing well out from the
head. Another sign of the aggressive phase of this faculty
is known by shaking of the head from side to side and forward and backward, while engaged in an energetic conversation. A short, low nose, with a high and thick pug end, is
evidence of pugnacity. All the noted prize-fighters whose
portraits I have observed have this description of nose, and
a very short, thick neck, with great muscular powers generally; but moral courage and resistance springs from an excess
of Conscientiousness, and is a mightier force than that combativeness or resistance which proceeds from muscular development merely. Veneration, shown by height of the nose,
lends to the character the ability to combat argument and
opinions. Every faculty has its own peculiar force, or mode
of expressing power. These different methods of showing
force must be analyzed by the reader, else confusion will
ensue and motives will not be comprehended fully.

CAUTIOUSNESS.

One of the most reliable evidences of Cautiousness is seen
in the long nose. Its location is admirably adapted to the
preservation of the body, presiding, as it does, over all the
functions of digestion, and guarding the avenues of approach
to the stomach by its keenness of scent, which soon detects
qualities of food unsuited to the sustentation of the body.
Short noses are not so efficient in guarding these functions
as long ones.

This faculty is general, and is useful in all the walks of life,
giving forethought and carefulness, making the individual

EDMUND SPENSER. (CAUTIOUSNESS.)

Cautiousness is one of the leading signs of character in the countenance of this portrait of Spenser, an English poet of the sixteenth century. His biographies attest that he was exceedingly cautious, almost to timidity. Delicacy, sensitiveness, and shyness are often the accompaniments of genius.

provident, discreet, and vigilant. It is large in all those animals who have to depend more upon their carefulness than strength for their subsistence and preservation. The fox, rabbit, hare, and squirrel all evince a high degree of this trait. It is larger in women than in men. Woman's conservative nature requires the exercise of its power for the protection of the family. Sly children are largely endowed with Caution. Its excess causes irresolution, melancholy, fear, terror, suspicion, cowardice, dementia, and suicide.

HOPE.

MADAME D'ARBLAY. (Hope.)

Hope shines pre-eminent in every feature of this lovely woman. Mme. d'Arblay possessed a mind as hopeful as it was original. Her vivacious style, as exhibited in her tale of "Evelina," as well as her face, attests the presence of the faculty of Hope.

A high development of this faculty is found large in those whose septum, or middle partition of the nose, projects downward below the alæ, or sides. The part where the septum joins the upper lip is devoted to the sign for this trait, although clearness of the skin and eyes is another facial indication of Hope. Where this is well defined, the liver will be naturally active, and, if properly used, will never be the subject of attacks of biliousness. Bright, wide open eyes, buoyant and elastic step and carriage, are also evidential of Hope. Drooping of the corners of the mouth, with dull eyes, denote a lack of this faculty.

ANALYSIS.

CANOVA.   (ANALYSIS.)

The local sign for Analysis is well defined in the accompanying physiognomy of Antonio Canova, a renowned Venitian sculptor of the eighteenth century. His analysis of character, as illustrated in his grand and sublime classic statues, challenges the admiration of the world.

Analysis, as shown by this portrait, illustrates the ability to analyze, classify, and suggest. It is related to architecture, mechanism, literature, human nature, music, and the laws of Nature generally. It endows its possessor with the capacity to suggest inventions, improvements in art, music, literature, etc. It is accompanied with a fertile, suggestive, criticizing mind, and is ever ready with expedients and resources. Its facial sign is known by a drooping of the septum of the nose, just forward of and adjoining Hope. Its action is affected by the liver, but not in the same degree as is Hope. The physiognomies of La Place, Dr. John Hunter, Dr. Jenner, Canova, the sculptor, Sarah Siddons, actress, and Arkwright, inventor, all exhibit this sign well defined.

MENTAL IMITATION.

JOHN GREENLEAF WHITTIER. (Mental Imitation.)

The sign for Mental Imitation is exhibited in the face of the Quaker poet, whose works seem to be imbued with the spirit of prophesy. They are all of an elevating tendency, evincing purity, sincerity, and beauty. His expressive countenance discloses all these traits, and many others as lovely. Imitation is very decided in the picture of Edmund Spenser, as well as in the faces of most authors.

Mental Imitation is located just under the tip of the nose, and forward of Analysis, causing a drooping of this part. It is found much developed in the faces of all persons in the imitative professions—in actors of the imitative class, in writers of fiction, in artists, poets, and in the faces of all persons in private life who exhibit marked imitative power in any direction. This sign is as indicative of mechanical imitation as of mental imitation, and those who possess a fine endowment of this faculty have the ability to copy the position, walk, voice, and gesture of others. A suitable physiological construction accompanies this sign, and all who have a large degree of Imitation have also a fine quality of muscle, which enables them to adapt themselves very readily to changes, and adopt the habits and customs of others. The Irish, as a rule, are very imitative, and early adopt the customs and habits of all countries where they find a home. The French are the reverse of this.

IDEALITY—TASTE—IMAGINATION.

SIR PETER LELY. (IDEALITY.)

Sir Peter Lely's nose discloses a large degree of Ideality, or love of the beautiful. He is noted for his fine landscapes and portraits of Court beauties, painted in England in the seventeenth century. His entire organism reveals the artist. Form, Size, Color, Constructiveness, Amativeness, and Love of Young, together with Imitation and Analysis, mark their local signs upon this expressive countenance.

There is no single term in the English language which expresses this faculty. Those possessing it in a large degree evince all three of these traits. It is known by width of the tip of the nose, often giving it a square cut appearance. It is noticeable in the faces of artists, poets, sculptors, actors, and certain classes of the literati. In private life, all persons who exhibit good taste and refinement have this indication in their countenances. Sharp pointed noses show just the reverse of this; such features denote very matter-of-fact, unimaginative characters. The portraits of Edmund Burke, Lord Byron, Washington Irving, Corneille, Akenside, Mrs. H. B. Stowe, Horace Vernet, and a host of other writers, artists, etc., exhibit this sign of character.

SUBLIMITY.

SIR JOHN HERSCHEL. (SUBLIMITY.)

Sir John Frederick William Herschel will ever be renowned for his grand discoveries in Astronomy. His sublime and far-reaching intellect recognized the interrelation and correspondence of all sciences. His vivid imagination assisted, by its comprehension of analogy, his great efforts in discovery. The signs for Sublimity and Ideality are wonderfully apparent.

7

The sign for the love and appreciation of the grand and sublime in Nature, art, and conduct of life lies just each side of and beyond Ideality, producing a full and rounded appearance of the end of the nose. It is large in those who excel in depicting the grand and sublime by pen, brush, or chisel. It is also seen in the faces of all who appreciate this quality in art or Nature, for faculties are both executive and appreciative. Some persons possess one form of this talent, and some the other, while others have both. Where this faculty is large, in conjunction with a good development of Acquisitiveness, it will lead its possessor to plan and execute large business enterprises and financial schemes. Persons who exhibit large Sublimity have broad and comprehensive views of life and of business, and never like to undertake anything on a small scale. They have either physical or mental power to carry out their plans. Jewish physiognomies have this sign large.

HUMAN NATURE.

Human Nature is found lying between and above Ideality, and fills out the nose at this point. In some it produces a decided turning up—what the French call the *nez retroussé;* and, as it indicates a knowledge of human nature, it gives with this formation a piquancy, aptness, and keenness in conversation and repartee, caused by a ready apprehension of the character of the person with whom one is conversing. It also gives an intuitive comprehension of Physiognomy and of truths as they relate to the human mind and body. Noses which rise high above the plane of the face at the tip indicate a far-seeing understanding of human nature and of all nature, combined with a love for its study and investigation. Examine the noses of Aristotle, Porta, Lavater, and Simms, in the front of the Introduction to this volume, for proof. It endows one with an insatiable desire to know the facts of Nature as they relate to Man. An individual with this faculty large can readily become a physician and hygienist.

JOHN LOCKE. (HUMAN NATURE.)

Human Nature large. John Locke, the celebrated English philosopher, is noted for his "Essay on the Human Understanding." It shows him to have had the ability to penetrate the inmost recesses of the human mind. His nose, rising high above the plane of the face at its tip, is a corroboration of his gift as a seer of Nature.

Another sign of Human Nature is found in the fullness of the ethmoid bone at the inner corner of the eye. This, joined with width between the eyes (Form), shows the power to retain the memory of faces, forms, etc.

Human Nature is very large in children, and some adults possess it in a large degree. Where the point of the nose stands high up from the face, a ready apprehension of human nature exists. Knowledge of human nature is the most essential to human happiness and welfare; hence, it occupies the most prominent position in the face. A good share of intelligence on this subject will lead to suitable selection for partners in marriage, and in this way it tends directly to race improvement. Doubtless, this is its primary use. The countenances of all eminent doctors, botanists,

naturalists, philosophers, and scientists exhibit this trait.
It is noticeable in pictures of Cicero, Hippocrates, Averroes,
and Avicenna, among the ancients; it is large in the faces of
Drs. Jenner and Hunter, Sir Astley Cooper, Corvisart, and
hundreds of similar characters.

CONSTRUCTIVENESS.

SAMUEL F. B. MORSE.   (Constructiveness.)

The inventor of the electric telegraph was, as his nose indicates, both mechanical
and artistic.   He excelled in both directions.   Boldness and originality are stamped
upon every feature.   Observation is also very marked.

Constructiveness, or mechanical ability, is known by fullness of the sides of the alæ, or wings of the nose. It joins
Human Nature and Sublimity on the upper side, and Acquisitiveness on the lower side. A good constructive nose is
well filled out at the sides below the bridge. Noses having
a depression or indentation at this part lack mechanism.
This faculty is needed in art, to assist the painter, sculptor,

novelist, actor, and musician, in order that they may be able to construct their work on mechanical principles, for without mechanism as a basis, art could not advance. Mechanical construction is as essential in planning a sermon, a novel, or a history, as it is in building a house. The faculties found in combination with Constructiveness will decide what particular thing will be constructed—whether a book, a statue, a painting, or a church.

ACQUISITIVENESS.

MICHAEL REESE. (ACQUISITIVENESS.)

The accompanying cut reveals the faculty of Acquisitiveness on a large scale and in many places. The signs in the droop of the eyelids, the thick straight nose, and dry appearance of the lower lip disclose the fact that Michael Reese, the capitalist of San Francisco, lived for the pleasure of accumulation purely. The entire organism shows the predominance of the accumulative principle.

This sign fills out the nose just where the alæ *join* the face. The "hooked nose," such as is often seen in Jewish

faces, is another unfailing indication of acquisitive power,
In those in whom this faculty is deficient a hollow is seen at
this place. This faculty gives the power to acquire property, learning, or power, or whatever the faculties in combination with it lead the individual to desire most. It is sometimes seen in the heavy eyelid, which shows the whole of the
upper lid while the eye is yet open. This feature is most
seen in Oriental faces. This indication is well illustrated in
portraits of Sir Moses Montefiore, Alexander T. Stewart,
and George Washington. Benjamin Franklin was large in
Acquisitiveness as well as economy, and was charged with
penuriousness. The combination of traits in his face sub-
stantiates the charge. When biographies of great men do
not agree with their physiognomies, I always believe the
latter in preference to the former.

VENERATION.

At the highest part of the nose, very commonly called the
"bridge," Veneration is located. The higher and broader
the nose at this point, the greater the development of Veneration. This faculty gives the power to venerate or respect
God, law, Nature, old age, antiquities, and every idea, person, and thing according to its merits. The most noted
divines and commanders of all ages have possed this faculty
largely. Those with Veneration large make good commanders, for those who can the best command can the most readily
obey; the faculty that gives one the power to comprehend
and respect law produces the desire to see it carried out,
either by executing or obeying its provisions. The ability
for logical argument is one attribute of this faculty, and
where the nose is broad as well as high at this sign large reasoning powers may be inferred, as well as capacity for logical
ratiocination. Plato, Huss, Wickliffe, Luther, Emanuel
Kant, Calvin, Fontenelle, the two Mathers, John Locke,
Benjamin Franklin, George Herbert, and Canon Kingsley,

REV. DR. MUHLENBURGH. (Veneration.)

Veneration is exhibited in this countenance of the Rev. Dr. Muhlenburgh in a most conspicuous manner. His life-work tells us how large a place it filled in his nature. He founded hospitals, homes for the aged, and retreats for crippled and destitute children, and lived a celibate life in order to carry out his plans for the needy of earth. His Veneration gave him respect for all that merited it—God, old age, the helpless and desolate.

all have this faculty in a remarkable degree. Veneration is not, as some imagine, a slavish adoration of a deity, or of a number of deities. That is the perversion of the true use of this noble faculty; or, rather, its use without reason. Veneration should always be moderated by reason, or subordinated to it; for, without it, it degenerates into many gross errors and unjust customs.

Lecky, in his "History of European Morals," says: "Reverence is one of those feelings which, in utilitarian systems, would occupy at best a very ambiguous position, for it is extremely questionable whether the great evils that have grown out of it in the form of religious superstition and political servitude have not made it a source of more unhappiness than happiness." This idea shows that it can be turned from its legitimate use, and work injury to whole nations. Still, no character is truly beautiful or noble without it.

Persons deficient in Veneration have a decided depression at this point. They are never leaders nor commanders, and do not yield ready obedience to law. Not apprehending the nature of veneration or respect, of course they cannot easily put it into operation. Where there is a great lack of Veneration, we are sure to find impudence and impertinence in excess. Negroes as a class have little reverence. Their religious feelings proceed almost entirely from an over-development of Credenciveness, making them grossly superstitious, as they have little reason to control this faculty. In all undeveloped people and races, an excess of Credenciveness is thought to be a religious trait, whereas it is only a mark of an unbalanced mind. True religion is shown where reason and morality hold the balance.

EXECUTIVENESS.

SIR CHARLES NAPIER. (EXECUTIVENESS.)

This leonine countenance announces to the most casual observer the qualities which characterize the king of beasts—power, force, and courage. Admiral Sir Charles was a descendant of a family noted for its warriors; hence, he inherited Executiveness in an intensified form. His physiognomy speaks for his character. No biography is needed here.

Executiveness is found on the ridge of the nose just above Veneration, and forms the outline generally called "Roman," from its having been the form of the nose of many of the leading Roman generals, commanders, orators, and statesmen. It is the faculty which gives ability to execute law and order. It makes the individual bold, resolute, and daring in the discharge of duty, with power to lead rather than to be led. It is found in the physiognomies of Hannibal, Julius Cæsar, Charlemagne, Duke of Marlborough, Lord Nelson, Marshal Saxe, Duke of Wellington, Generals Scott and Sherman, and all who have become known to history for their abilities to command and lead large bodies of men in aggressive movements. With such men it assumes the aggressive form, and peculiarly adapts them to fill the position of military leaders. Unaccompanied by a strong sense of justice, it becomes tyranny.

SELF-WILL.

Self-will is located at the root of the nose just between the eyes, and is indicated by height at this part. It is found more largely developed in European noses than in American, and it is this which gives the plodding, persevering character so often observed among them. Americans, as a class, do not continue and dwell so long at their undertakings as Europeans, and, consequently, show a greater depression in the descent from the forehead to the nose. Self-will is one of the most important faculties in the human mind. It is found small in undeveloped races and faces, and large in all who have excelled in any work requiring continued attention, as have the following named persons: Dante, Josephus, Sir Isaac Newton, Humboldt, Milton, Omar Pasha, John Stevenson, Richard Arkwright, Kossuth, Mrs. Harriet Beecher Stowe, Mme. de Stael, Edwin Booth, Brigham Young, Cyrus W. Field, Horace Greeley, and C. P. Huntington. Female physiognomies are seldom seen with this sign large. Their

COUNT VON MOLTKE.   (Self-will.)

Self-will is stamped on every lineament of General Count Von Moltke, eminent alike as a commander, strategist, and writer. His strong will, in combination with his intellect, led him to success in vast military enterprises.

occupations preclude its cultivation.   Woman's life is made up of trivialities and constant changes of small occupations. Very few have the opportunity to pursue one grand object in life.

Some observers give as one sign for this faculty horizontal wrinkles across the top of the nose, where it joins the forehead.   My observation affirms this sign, but it is found only in adults after concentrated effort in a given direction has been made long enough to stamp its effect upon the muscles at this point.   I have never seen this sign in childhood. The former sign of height where the nose joins the forehead is indicated in children, as well as in adults, as showing the ability for consecutive efforts of the will.   Those persons who have been successful in carrying forward great enterprises—such as building railroads, laying telegraph cables, etc., founding and managing communities, and in all undertakings in art and invention requiring persistent Self-will— show this height of the nose. This faculty occupies a commanding position in the face, and thus shows its importance to the intellectual operations, as well as to the social and domestic faculties. In most faces in which it is predominant the sign for Firmness is often small; the chin will be found retreating somewhat. In rare instances you will find both these faculties large, and this will give an exceedingly willful, obdurate, set character. Herder says of this locality: "The region of the face where the mutual relations between the the eyebrows, the eyes, and the nose are collected, is the seat of the soul's expression in the countenance; that is, the expression of the will and of the active life." Europeans use the word *soul* often to designate *mind;* I think this is the sense in which Herder has used it in this passage.

When Self-will is large, the Muscular will be one of the dominant systems of the individual. Will is operated by the muscles. Firmness, which is erroneously called will-power, is evolved from the Bony structure and operated by it. When one analyzes the constituents and nature of bone and muscle he will observe the propriety of this distinction. Will is very changeable like muscle; Firmness is immovable like bone; hence the difference in their operation.

CREDENCIVENESS.

This trait has been named both Spirituality and Faith, neither of which expresses its real office. It gives a love of history and belief in tradition and biography. Perverted, or unbalanced by reason, it leads to superstitious belief in all sorts of improbable dogmas and wonderful or impossible stories. It is located at the inner corner of the eyebrow, where it joins the nose, causing an arched appearance of the eyebrow at this part. It is found large in the Chinese, in Turks, in Jews, and in all Oriental nations, who have for

JOHN DE WICKLIFFE.   (CREDENCIVENESS.)

This eminent divine of the English Church made a determined stand against the errors of the Romish Church, but was himself a believer in many things not consistent with reason or the truths of Nature.   He saw only the "mote" in his brother's eye, not the "beam" in his own.   He lived a moral life, however, and was popular in the fourteenth century.

ages listened to the recital of superstitions or wonderful traditions.   It is shown, also, by the wide open eye, and is small in scientists and those who are so constituted as to demand proof before belief.   Its absence in them is shown by an appearance just the opposite of this, or by eyebrows drawn close down to the eye, thus bringing the eyebrow down so close as to make the eye seem small.   Where it is uncommonly developed, it begets a love of the wonderful and superstitious.   The individual will place implicit reliance on all great and improbable narrations, such as relate to ghosts, spirits, and great snake stories; and many persons, of good judgment in all the affairs of every day life, will accept as a religious belief statements founded on the impossible, and which reason and truth show to be so.

This faculty is very general, and is found in degrees ranging all the way from its legitimate action—viz., a belief in history, biography, and well authenticated facts—to childish credulity and belief in the impossible—in fairies, ghosts, genii, dragons, and in the religious power of charms, incantations, and in the sanctity of certain animals, birds, and insects. "Plain sense will influence half a score of people at most, while mystery will lead millions by the nose," said Lord Bolingbroke.

The sign of this faculty is found excessive in the faces of Joan of Arc, Bajazet, Ignatius Loyola, Schamyl, a prophet and military commander of the Circassians, Pope Alexander, Mahomet, and Swedenborg, and other "believers."

OBSERVATION.

CICERO. (Observation.)

Marcus Tullius Cicero, Roman orator, statesman, and jurist, discloses by his facial record that Observation was his natural and cultivated talent. The constant use of the muscles at the place where the inner termini of the eyebrows approach, together with the tension of the muscles of the eyes toward Observation, contribute to form the protuberance found in the face of this great observer and thinker.

Observation is a faculty of the Muscular system, and is related to the observation of all visible objects as separate and individual existences and entities. It gives an insatiable desire to look at and examine everything, to ask the use and purpose of every object in sight. It is always large in children, and the faculty most used by them, and through which nearly all knowledge comes to them in the early years of life, before the practical faculties and reason have begun to assert themselves. It should always be cultivated in children; never repressed by avoiding answering their questions or forbidding them to gaze about. Its situation is in the middle of the lower part of the forehead, between the inner termini of the eyebrows and above the top of the nose; when large, filling out the forehead to a level with the adjoining parts, and causing the eyebrows in adults to lower down toward the eyes at their inner corners. This last sign is not observable in children, the faculty not having been put in use sufficiently long to leave its impress upon the muscles. It will be found large in scientists and mechanics generally.

It is the opposite of Credenciveness, and its sign is also the reverse of that faculty. It depends upon personal observation for its proof of all things, and, therefore, is small in theologians and the superstitious, who accept doctrines unproved, upon faith and without demonstration. The portraits of Arago, Buffon, Michael Angelo, Fulton, James Watt, Professor Tyndall, Charles Darwin, Cyrus W. Field, Professor Morse, Elihu Burritt, John and Joseph Le Conte, and Dr. J. Simms show Observation very well defined.

FORM.

Form is that faculty which gives the power to recollect configuration generally. The memory of faces, figures, pictures, spelling, distances, and length proceed from a development of this faculty. It is large in draughtsmen, artists,

WILLIAM HOGARTH. (FORM.)

The faculty of Form shines pre-eminent in the impressive face of Hogarth, the most eminent of all caricaturists. His etchings and engravings have a world-wide renown. Width of the bony structure between the eyes is the chief facial sign of this faculty. His countenance discloses mirthfulness without stint. His sense of the ludicrous and grotesque is well illustrated in his "Marriage a la Mode" and "The Rake's Progress." He lived in England in the seventeenth century.

wood engravers, wood carvers, sculptors, etc. It is exhibited by width between the eyes. Persons with large Constructiveness, or mechanical ability, also retain memory of form well. Scientists usually possess it in a great degree. It is useful in classifying and arranging physical substances, and is needed in Architecture, Geometry, Botany, and Comparative Anatomy. Dressmakers and pattern-makers must possess this faculty, and all who employ form as the basis of their trade or profession.

It is found large in actors and elocutionists, who excel in gesture. Kean, Siddons, Rachel, Madame Ristori, Booth, and all those who have excelled in delineating the posture and form of others, show this faculty in excess. It is uncommonly large in the faces of Cuvier, the naturalist; Linnæus, the botanist; Baron Humboldt; Sir Astley Cooper, the celebrated surgeon; Watt, Fulton, and Stephenson, who were mechanics and inventors; and in all the great artists. "An artist," said Michael Angelo, "should have his measuring tools not in the hands, but in the eye." Form, as well as Size, are caused by the peculiar conformation of the Bony structure.

SIZE.

See portrait of Hogarth at Form.

Measurement, by the eye and memory, of distances, proportion, and magnitude, are all the result of a development of the faculty of Size. Persons endowed with this faculty, in conjunction with Form, make good architects, and can estimate accurately lengths, widths, angles, quantities, and centres. Its location is on either side of Observation, and gives a V-shaped appearance to the forehead just above where it joins the nose.

Hogarth is a fine illustration of both Form and Size; also, Rembrandt, Durer, Van Dyck, Gerome, and all persons renowned for their power in comprehending and delineating the size of objects. Physiognomists possess this faculty as well as Form. The faces of Aristotle, Porta, Lavater, Dr. Gall, the phrenologist, and Dr. Simms have these last two signs very large. Surveyors, engineers, architects, and inventors have Size and Form well developed. In the face of Brunel, the architect of the Thames Tunnel, it is remarkable. Persons whose eyes are placed close to the nose, with little width between, are deficient in Size. Many otherwise well educated men cannot spell correctly, owing to the absence of Form and Size.

WEIGHT.

Weight gives an intuitive appreciation of the laws of motion, balance, and weight. Persons with a good endow-

ARAGO. (Weight.)

In this "counterfeit presentment" we have the local sign for Weight very large. Arago, a distinguished astronomer, mathematician, and scientist of France, proved by his discoveries that the faculty of Weight, or Balance, was largely represented in his organization. The V-shape at the termini of the inner ends of the eyebrows is one of the facial indices of this muscular faculty.

ment of this faculty are natural climbers, skaters, dancers, etc. It is always found large in rope dancers, billiard players, marksmen, gunners, and archers. It is an essential power in carpenters, architects, and sculptors, as it enables them to plumb lines by the eye, and gives a correct and natural pose to the figure. The architect of the Leaning Tower of Pisa must have possessed this faculty in an eminent degree. I have observed it in acrobats, gymnasts, trapeze performers, velocipedists, and sailors, and in all who require good balancing power.

The sense of Weight might properly be called the sixth sense, as it enables us to balance in walking, and to judge of the weight of articles by lifting. It is related to the Muscular system; therefore, the shape of the hands and feet

8

would indicate its power. Its location is adjoining Observation and Credenciveness, and is a faculty that carpenters, iron workers, and astronomers must possess in order to excel in their several callings. When it has been much exercised, the muscles called *corrugator supercilii* increase in size, and this is another proof of its muscular origin.

This sign is noticeable in the portraits of Sir Isaac Newton, the Herschels, R. A. Proctor, Brunel, the architect, and Blondin, who crossed Niagara Falls on a wire. It is adapted to muscular control, to the use of tools, machinery, to swimming, dancing, etc. George Combe, the phrenologist, states that those in whom it is largest are least subject to sea-sickness, which proceeds from two causes: Lack of balancing power—that is, absence of the sense of weight; and, second, from an abnormal bilious condition of the liver and stomach. The muscles and bones assist in forming a system of levers and other mechanical powers. A predominance of these two systems—the Bony and Muscular—contributes to the mechanical powers of the individual. Now, mechanism of all kinds—whether it be by the voice, brush, pen, or chisel; whether by skillful manipulation with the hands on the piano or sewing machine, or by practiced movements of the hands and feet—is indebted to the muscles, principally, for the illustration of its power; hence, we must look to the degree of muscular development of the entire organism for artistic and mechanical power. The more bone, the more mechanical, exact, and scientific; the more muscle, the more artistic, emotional, and ideal.

Singers, like Jenny Lind or Grisi, depend upon the fine quality and rich endowment of the muscles of the larynx and ear for their powers of song. The orator, also, finds in the resonant quality of muscle his sure ally; while the mechanic and sculptor must depend upon bone, muscle, and nerve for the keen sensibilities of their hands to perfect their useful and artistic creations. The perfection of vocal expression, whether in song, oratory, the drama, or in language, is reached when the Muscular system is of fine quality and abounding. The sculptor must be able, by his natural and acquired sense of weight, to judge how to graduate the blows upon his chisel in order to produce the effects which he desires upon his marble. His ability is the result of an organization suited to his work by nature; then perfected by cultivation and use.

Thorwaldsen, Canova, and Miss Hosmer represent in their physiques a strong endowment of the Muscular system and sense. The sign for Weight is uncommonly large in all these great artists, and their hands are the hands of artists. Any observer, after mastering the principles underlying the human organism, as laid down in this work, should be able to give most of the salient points of character from the hands alone.

COLOR.

A fine perception of Color is shown, generally, by a decided color of the eyes, skin, and hair; that is to say, the deeper the colors in one's organism, the better the sense and appreciation of the harmony of shades and tints. Persons with light hair, eyes, and skin are not so capable of becoming good colorists as those with darker hues in their physiognomy, and this principle of Nature is obvious. We can give out only that which we possess, and those who are deficient in color cannot perceive its shadings and harmonies so well as those whose natures are permeated with it. Color is also shown in the arching of the centre of the eyebrow, as seen in the face of Guido Reni.

Color is of far more importance in the economy of Nature than is generally known. Recent discoveries in the principles of color have shown its power as a remedial agent, and colors are now applied to various diseases with great success. Color is an effect of climate, soil, food, and drink, acted upon by the solar rays. Every mineral which enters into the earth's composition, or into that of the sun, gives forth its color, which has a perceptible and powerful effect upon the

GUIDO RENI. (COLOR.)

The portraits by this distinguished artist are compared to those by Raffaelle in the brilliancy of their coloring. Dense color pervaded his entire body. Olive complexion, red cheeks, and black hair and eyes prepared him to comprehend the harmonies of colors, shades, and tints. Lavater says, "We read the coloring of Guido in his countenance." This portrait represents the artist at an advanced age.

human organism. These rays must harmonize with the conditions of the human system in order to produce health. Many persons are color-blind. A very much larger percentage of men than women are so. A depression or lowering of the eyebrow at the centre is an infallible indication of colorblindness. A recent writer on color-blindness, Dr. Jay Jeffries, estimates the percentage of those who are colorblind, in a greater or less degree, as one in every twenty-five males. The proportion is much less in females; Dr. Holmgren found but ten women in seven thousand one hundred and nineteen, of all ages and occupations, color-blind! The reason why men are more deficient in the sense of color is that they do not make the same study and use of colors that women do, in their dress, and in the pursuit of their tastes generally; in the disposition of house and flower decorations, and in adapting colors to complexion. The use of tobacco and alcohol is thought to injure the color sense. The principle that "use increases capacity" applies as well to this department of the organism as to all the other parts of the system. Professor Holmgren says that "color-blindness is not a disease in the sense of being attended with suffering, obliging the person to have recourse to a physician. Color-blindness, quite as well as normal sight, is a sense of color, though of another and more simple nature. He whom we call color-blind is not, correctly speaking, at all blind to all colors. In the system according to which he arranges his colors, he has *fewer* kinds than the normal observer. It results from this that he finds resemblances between colors, or confuses others that the normal observer finds different; for instance, red and green."

There are many unsatisfactory theories put forth to account for the deficiency of the color sense. My own ideas on the subject may be useful as far as they go; I know, however, that they do not cover the whole ground. My observations have led me to remark two causes for this defect: First, the lack of foods which contain those elements that produce the kind and amount of color essential to the healthy equilibrium of the organism; that is to say, that in the chemical combination of the food with the blood and tissues, there is not sufficient coloring matter mingled to endow the person with the right proportion of color to constitute a strong and decided color sense; also, there is not enough of color derived from the solar rays. This proceeds from a disregard of sanitary law in pursuing an in-door existence, or a non-assimilation by the organism of these rays, in consequence of certain diseased conditions which prevent, for a time, the proper action of the light and heat of the sun. This opposition is, however, only temporary, as it is well known that sunlight alone will eradicate many diseases; and, as the white rays of the sun are composed of a combination of all the colors of the prism, the curative properties must reside in the colors alone. If this were not the case, a heated room would conduce to health as well as sunlight. Experience proves that this result cannot be obtained without the direct rays of the sun. Plants languish and become pale and sickly when deprived of sunlight, and vegetable juices undergo serious chemical changes from being shut off from the action of the solar rays. There are other sources of light and color which are nearly the same in their composition and action as sunlight. Electricity is one of these sources which has a direct bearing upon the health of organic life; and, although many of the laws relating to this force are unknown, still enough of its action has been observed to assure us that a proper amount must enter into the constitution of the human organism to produce healthful conditions. There are other sources of light, which cannot be dwelt upon here. The reader might consult the works of Descartes, Helmholz, Huyghens, and Draper on this subject with profit.

The second cause of defective color sense is plainly proved, I think, by the fact that a much larger percentage of males are lacking in this direction than females. The reason for this has been previously stated; viz., non-use of colors habitually. Colors are interwoven, one may say, into the every-day life of woman, while only a few men, comparatively, pursue callings into which colors enter as a prime factor. The theory of non-assimilation in the organism of mankind, by chemical action, of sufficient color to give a correct and just understanding of colors, should teach us how important, in a moral sense, a due development of color is to the human body. The reader will observe, in the chapter on sub-basilar principles, the reference made to color by the celebrated naturalist, Haeckel, who has observed that the absence of color induces or accompanies abnormal conditions, both in animals and man. He, however, gives no theory on the subject.

Absence of color produces not only physical defects—as, for instance, the absence of the color sense—but also moral deficiencies. Now, very light gray eyes, or almost all light eyes, are indicative of either scrofulous tendencies or weakness of the kidney system; and weakness or deficiency of that system shows a lack of natural integrity or Conscientiousness. I know that this theory will be assailed and scouted. Let me ask my opposers to bring as good proof against this theory as I can to its support, before they proceed to condemn it. Surely, twenty-five years of observation in the study of the human organism have not failed in discovering what the eye of the indifferent would never become cognizant of. At the same time, I would advise all doubters to take heed of Descartes' maxim, "Assent to no proposition the matter of which is not so clear that it cannot be doubted." As the moral as well as mental powers depend upon the constitution of the atoms and molecules which compose the cellular tissues of the body, how can it be expected that integrity shall be one of its components if chemical action has failed, in the first instance, to properly blend and harmoniously balance the physical organism? Morality is not a fine spun, fleecy, cloudy theory of *belief;* Conscientiousness is not an intellectual opinion as to the merits of Infant Damnation or the doctrine of Salvation and Election, or any other purely speculative belief. It is the very groundwork of our physical construction; it inheres in the chemical or underlying basis of our organism, and depends for its soundness on the purity of the body primarily, and afterward on a cultivated and quickened moral sense.

A more general comprehension of colors would lead to greater protection of life and property. Colored signals and lights are used largely on railways and on shipboard. The failure to understand the colors of these signals might prove disastrous in the highest degree. Men intending to enter positions where such signals are used should be subjected to an examination in colors before assuming such responsibilities, where ignorance of colors might jeopardize the lives of hundreds. The Color sense is dependent upon the normal condition of the glandular system and the arterial circulation for its illustration. The local sign (arching of the centre of the eyebrow) is an inherited feature, transmitted from ancestors who have had the Color sense cultivated in trades or professions, and it then becomes a permanent formation. This sign is also caused by the roundness and size of the muscles of the eye and its orbit; also, by raising the lid and brow constantly, as all artists do in their work.

ORDER.

CUVIER.   (Order.)

Baron Cuvier's brow illustrates the local sign for this faculty admirably. His classification of scientific subjects, and his collection of natural objects, aroused this sense to its highest activity. He is remarkable among the greatest scientists for his systems of classification and orderly arrangement. The Bony system is one of the ruling powers in him.

The next local sign in the Architectural division is found externally from Color, and adjoins Calculation, its natural assistant. When very large it forms almost an angle, and assists in forming the arch produced by the local sign for Color. It is adapted to method, system, regulation, law, and custom, and is essential in business, housekeeping, and in all the trades. It is found large in apothecaries, printers, accountants, scientists, and librarians. Those with a large share of Order are pained by disorder, and must have a place for everything and everything in its place. Order derives its power from the size and form of the Osseous system. The bones of the forehead form the shape which is assigned to the local sign for Order. The phrenological idea, that it is an *organ* of the brain, is wholly false. It inheres in the squareness of the Bony system, and is not found in the Vegetative along with fat. It belongs to the *mind of the bones*.

CALCULATION.

GEORGE BOOLE. (CALCULATION.)

Calculation very large. The square bony frame of George Boole, in combination with the muscular powers, enabled him by its orderly arrangement to calculate uncommonly well. His power of reason added made him one of the most profound mathematicians of the world. His works are too abstruse for general use; hence, he is little known.

The ability to calculate numbers, reckon figures in the head, and remember figures, dates, and sums, is shown in the face by width of space between the external angle of the eye and the eyebrows, as shown in the above portrait of Boole. This faculty varies in different individuals in an astonishing degree. While some can compute immense sums almost instantaneously, others, who are exceedingly intelligent on many other subjects, are often nearly idiotic in this. George Combe said of himself that, after seven years of study of arithmetic, the multiplication table was a profound mystery to him. Calculation is found more largely developed in the Chinese than in any other race. The teacher of a Chinese mission school in San Francisco informed me that young Chinese boys would commence with notation, and in four months could calculate division of fractions. Two years are required to carry the white children of the public schools through the same course. Individuals possessed of Arithmetical Calculation, in conjunction with Form, Size, Causality, Comparison, and Constructiveness, can become both mathematicians and geometricians; with Locality and Imitation added, can easily become surveyors and civil engineers. The physiognomies of La Place, Herschel, Lagrange, Brindley, Stevenson, Smeaton, Bradley, Lalande, Gibbon, and Delambre, and all others who have excelled in astronomy, history, physics, surveying, engineering, and similar pursuits, express the faculty of Calculation. This faculty inheres in the Muscular system mainly, assisted by the Bony. We know this by the fact that those who have excelled in Calculation have possessed a good share of these two systems. We cannot go beyond or outside of Nature for proof of natural operations. Those in whom the Vegetative system is regnant are very poor calculators, have little Order or Locality, and are deficient in the Bony and Muscular powers generally.

LOCALITY.

Localizing sense gives the power to place everything in existence as seen. It also gives the memory of scenery, cities,

CHARLES ALFRED TOWNSEND. (LOCALITY.)

A talented journalist, well known for his accurate and brilliant descriptions of localities and persons.

roads, and all places, things, and localities once viewed. All scientists have this faculty large. It is indispensable to the naturalist, geographer, astronomer, chemist, and descriptive writer, and in all the practical industries. Noted travelers and navigators possess this faculty in a large degree, and all persons who find and remember places, who can localize minute objects in the home, store, or factory, are indebted to this trait for their capacity in this direction. Locality, in combination with Constructiveness, gives the ability to remember machinery and set it in position; with Observation, Human Nature, Memory of Events, and Form large, a desire to travel and to study men, things, and the world generally is evinced. It is found just above the local sign for Weight, running upward and outward, and is so large in some cases as to resemble a small wen or kernel. The study of anatomy, physiology, and physiognomy, as well as of all other sciences, cultivates and develops this power. The sign for this faculty is strongly defined in the faces of La Place and Herschel, the astronomers; Sir Martin Frobisher, Sir John Franklin, Captains Ross, Cook, and Parry, navigators; Houdin, the French conjuror; Marco Polo, traveler; General Fremont, explorer, and Sir Walter Scott, writer. It derives its power from the Muscular system, and is largest in those who are given to excess of motion.

In giving local signs, it must not be understood by the reader that the faculty or power is located with the sign. The faculty and power are general, are diffused through the individual, and inhere in the entire system which the local sign indicates it as belonging to; as, for example, Weight belongs to the Muscular system, Constructiveness to the Muscular, and Form and Size depend upon the conformation of the Bony system.

MEMORY OF EVENTS.

The Memory of Events, as its name indicates, gives the power to retain and recall events of all kinds—history, scientific facts, anecdotes, experiments, public measures, and news and neighborhood gossip. It is located above Observation, and between the two local signs of Locality. Those with this faculty large readily learn and accept new doctrines, principles, and ideas; can become teachers, and, with Language large, editors and writers. It endows the individual with a common sense view of affairs and assists progressive tendencies. It is large in children, as their faces indicate. It is possessed by historians, descriptive writers, orators, and statesmen. The portraits of Victor Hugo, Dean Swift, Prescott the historian, Gladstone, and James Parton, all present this indication. Memory of Events is indebted to the Muscular system mainly for its power, assisted by the Osseous and Nervous systems. Its complex derivation gives it power

LORD CHANCELLOR SOMERS. (Memory of Events.)

Memory of Events is very pronounced in Lord Chancellor Somers, an English jurist of eminence. His numerous writings on legal and political subjects evidence his uncommon memory for events. His countenance, shown above, discloses by the fullness of the centre of the forehead that his works and physiognomy corroborate each other.

to remember events which the visual organs, the eyes, take cognizance of, as well as what is heard—as events transpire in history, or in affairs of the city, town, or neighborhood. The region about the eyes being well developed gives great practical inclinations to the individual, in all of which the eyes assist. In *listening* to news, the ear and auditory nerve are concerned, and thus this department of the Memory of Events is indebted to the Nervous system as well as to the Muscular. Memory pertains to every individual thing and fact in existence. There is memory of forms, words, tunes, time, taste, color, locality, numbers, faces, gesture, walk, motion, and of all things in existence. Memory is a faculty of each of the five systems of functions; each remembers its own sort of activity.

MUSIC.

The genius, love, and ability for music are shown in many ways and localities. It is more general and of the highest quality where the Muscular system predominates, with a due admixture of the Vegetative to give softness to its expression. It is less marked in the Bony system predominant than in the others. The reason of this is that muscle has the qualities of both resonance and contraction, which bone has not, being unyielding in its nature, and therefore not affording as much assistance to the sound waves. Two of the elements of sound are elasticity and resonance; therefore, air, strings, and wires convey sound best. Another element of sound is its wave, or curved motion; and, therefore, muscle is best adapted to the giving forth and receiving of sound. The larynx, where sound is originated in the human voice, is composed of muscles and ligaments; the bones have no part in the act of emitting sound. The sub-basilar principle, that "Nature is harmonious," is nowhere more beautifully exemplified than in the most prominent local sign for Music —the ear. Under this principle, the external must agree with the internal; therefore, if the body is built on musical principles, the brain must correspond of course, and the ear, being the channel through which sound is conveyed to the brain, must be adapted in its size, form, and color to the reception and conveyance of musical and other sounds to the tympanum.

Dr. Simms has given the best exposition of the *rationale* of this faculty of any writer on the subject. He says:

"The round or rounding ear, well set out from the head, arrests more of the passing musical waves, and conveys them with more force to the tympanum, or drum of the ear. This not only gives more power of hearing, but it conduces to musical judgment; for the organ, being put more on stretch, vibrates more rapidly to the nearly instantaneous succession of musical waves than the flabby ear that lies upon the side of the head, and, sponge-like, absorbs the force of the ear-

ARTHUR SULLIVAN. (Music.)

The round ear, round nose, and rounding muscular body of Arthur Sullivan admirably display the faculty of Music, as it inheres in the Muscular system. Musical talent was inherited by this gentleman, and has been cultivated thoroughly. He is a popular English composer, author of "Pinafore" and many other well known compositions.

vibrations, instead of conveying them to the internal part of the organ; hence, flat ears are not only unmusical, but generally dull of hearing. To make a good musician, the two ears must be alike, both of them being round, thin, and standing well out from the head, especially the outer rim. Musical ears are likewise generally red, because the vibratory motion with which they so readily respond to the waves of sound draws the blood to the surface of the organ. A bloodless ear is not one generally that has had much musical education. Another notable circumstance is that where the ear is largest at the top, the upper part being round, the person is capable of learning instrumental music. On the other hand, ears round at the top, but wide in the lower portion, with heavy pendant lobules, evince the faculty of vocal music; while width in the centre denotes the inclination to speak, and the power to speak well, so far as the tone and modulation of the voice are concerned.

"The musical ear in some cases is, and in others is not, a matter of inheritance. If the father or his relatives are musical, and the mother's family destitute of this talent, the children may be musical or otherwise, according to the parent that each of them represents in the musical heritage. When the two ears in any person are not alike in size and form, the right ear resembles either the father or the father's family, and represents their musical capabilities, while the left ear will be found like that of the mother or her near relatives. Sometimes it happens that the right or male side of the child represents a musical father, and that ear is round; while the left ear may be square, indented, angular, and broken at the outer edge, or depressed and flat, showing that the mother or her family must have been unmusical. On the other hand, if the father or his ancestors were deficient in musical talent, the rim of the right ear of the child will be more or less angular, or exhibiting straight lines, sharp turns or other deviations from the round form, evincing that mechanical rather than musical talent has been inherited from the father's side of the house. Some children inherit the musical as well as other capacities from ancestors two or three generations back. All round-eared animals love music, while long-eared animals are indifferent to or dislike it."

The Muscular system is mainly instrumental in producing sound.

LANGUAGE.

Language, or the ability to communicate by speech, has its signs in several places. The large, full, wide-open eye is an unmistakable evidence; so, also, a large mouth is proof of volume of language, and capacity, if the nostrils be large, for producing very loud tones. Full lips and throat are other evidences of this faculty. Many eyes appear small to the

LETITIA E. LANDON. (LANGUAGE.)

Language large. The fullness of the eyes and lips, and general roundness of the physique, show the predominance of the Muscular system. It is this system from which Language is derived. The numerous works of this lovely English poetess evince her great copiousness and power of language.

superficial observer because the bony ridge over the eyes projects so far forward as to hide their size, and very keen observation is needed here to distinguish their real size. One with large, full eyes and large mouth is fluent, voluble, and able to express himself with copiousness, freedom, and power; will be able to learn foreign languages easily; will possess great memory of language, and, with large reasoning powers, Causality, Comparison, Memory of Events, and large Observation, will have great power for oratory of high order. All vocal expression is dependent upon the Muscular system for its power. The vocal cords are cartilaginous, of the nature of muscle. The tongue, the mouth, and the larynx are mainly muscular. The ear, which receives sound, is indebted to the muscular sense for its effectiveness, and the eye, which "speaks," is also a mass of muscles.

9

The nation which reached the highest perfection in oratory was the Greek. This nation sought the perfection of the human form by the encouragement and exercise of those games which tended to the highest development of the Muscular system. The Olympian and Isthmian games were national and universal throughout ancient Greece, and were maintained at the expense of the government. They were considered sacred by the people.

One fact in connection with language, which philologists have overlooked, is that the language of races has progressed only as their physical powers have developed. The lowest races existing at the present time are as undeveloped in their muscular conformation as in their language. By tracing races physiologically, it will be found that their language has improved in the ratio of their muscular development. Language is the natural expression of the intellectual powers, and is also the channel of communication for our physical necessities. It is related to the three divisions of the face, and is necessary alike for physical, mechanical, artistic, religious, and mathematical expression. I believe it to be related to pulsation. Inasmuch as language is naturally and necessarily divided into pauses, there may be some connection between the beats of the pulse and the natural accentuation and periodicity of syllables and sentences. I give this as a suggestion to observers, not having investigated it sufficiently to state it as a fact in Nature.

The explanation of the locality and office of the faculty of Language completes the description of the Architectural group, occupying the middle portion of the face, and including the mechanical, artistic, religious, literary, and executive faculties.

THE MATHEMATICAL DIVISION.

All of the faculties and powers which relate to or assist in mathematical computation and demonstration are found in the upper or third division of the face, as exhibited in the outline cut on page 22. This attribute pervades all things, and shows the divisibility of substances, space, and time. Mankind would be like the blind groping in daylight without this power of computing, numbering, and demonstrating the numerical divisibility of all things in Nature. Statistics, surveying, navigation, weighing, measuring, and all business transactions involving calculation and accounting, come under the action of this department of the mind. Time in music, rhythm in poetry, the periodicity and revolution of the heavenly bodies, the succession of the seasons, and the quantitative particles of matter, are all subject to the laws of mathematics. So much of one element, another quantity of a different constituent, and a third proportion of some other substance, gas, acid, or ether, go to form every atom of organized life or matter of which the senses can take cognizance. There is no doubt that the pulsations of the heart and the natural accentuation of speech are subject to mathematical law. Indeed, there seems to be a law of correspondence throughout all Nature, by which the laws of all departments are correlated and act in unison with each other.

If the motions of the planets and the duration of the seasons, with all their sequences, are subjects of mathematical law and demonstration; if, in short, every atom of every kind whatever is regulated and governed by this all-pervading law of numerical certainty, why is it not reasonable to conclude that man's life, its duration and pathway or orbit through time and space, are also matters of law, coming naturally and necessarily under the law of mathematical certainty, and susceptible of demonstration, like every other atom, or organization of atoms, in the universe? You may say that this is but a restatement of what is called "the law of destiny." I do not object to that term if it be so understood as to include scientific law as the basis of the destination of all created matter. I do not give out this idea as based on a settled law of Nature, because I cannot substantiate it by well demonstrated facts; but reasoning from all the analogies of Nature, from the harmony that I observe attending all her operations, and from the co-ordination of all her forces, I believe that mathematical law may be the basis of the duration of our allotted time here. Its universality of application is simply unlimited.

"It is a character of all the higher laws of Nature," says Sir John Herschel, "to assume the form of a precise quantitative statement. The law of gravitation expresses the exact mathematical decrease of the gravitating force with the increase of the distances. Chemistry is, in a most prominent degree, a science of quantity. Astronomy likewise builds on mathematically expressed relations: the satellite revolving around its primary describes equal areas in equal times, and the squares of these periodic times are as the cubes of the distance. In the vegetable kingdom, two is the number ruling in the flowerless plants, three in the endogenous, and five in the exogenous. There is a mathematical law also governing the relative number of petals, sepals, and stamens, and the growth of leaves around the stalk. In animal life, the mollusk forms a perfect geometrical curve, and proportions the size of its whorls to the distance between them; and in the higher animals it is discovered, as in the number and size of the vertebræ, the number of teeth, etc., the same fact of a quantitative principle prevailing everywhere while yet in subordination to special laws of function or mode of life."

In commencing the examination of the Mathematical division of the face, I shall refer first to Time.

TIME.

Two kinds of ability to comprehend Time may be observed in the physiques and faces of George Boole and Arthur Sullivan, on pages 113 and 119.

Time is the faculty which enables us to take cognizance of the lapse of time, of periods of succession, of hours, days, months, and years. It is found large in the organisms of watch and chronometer makers, in astronomers, chemists, dancers, elocutionists, in the higher classes of mechanics, and in musicians. It is manifested in a variety of ways and by several local signs. As there are different kinds of this faculty in the human mind, it is logical to infer that it would be exhibited in the action of the different systems. This is in reality the case. When we find Time shown by persons of the round form equally with those of the square build, and then again another manifestation of this sense by those with the Brain system predominant, we conclude that its origin is as diverse as its manifestations are various and unlike. Many eminent musicians, like Handel and Meyerbeer, who were muscular and round men, illustrate the sort of Time used in musical exercises. It is large also in many mechanics and physicists who unite the Muscular with the Bony system slightly predominant. It is never found large in the Vegetative system.

Time gives the power to tell instinctively the hour of any given time of day or night, and ability to keep time in music. Persons with Time large evince a desire to be punctual, to have a set time for every act; can remember the dates when certain events transpired; will excel in keeping time in music and in marching. Time inheres in the entire organism—less probably in the Vegetative system than in the others, for in low animal organisms assimilation is constantly going on, but when we ascend to the higher, to the Thoracic, the periodic movements of the lungs and heart prove that this form of Time is expressed here. Although in that portion of the process of digestion carried on by the intestinal system there is a set time for the completion of the process of assimilation of the foods taken into the stomach as nutrition, in the Muscular and Bony systems we have still further assistants to periodic movements covering lapse of time; still in the Muscular and Bony systems we have more power evinced in this direction, in marching, in music, and in making instruments which illustrate the periodic movements of the earth, winds, tides, etc. Coming higher up, we find in the Brain and Nerve system the ability to comprehend the *lapse* of time, to compute its periodicity by mathematical calculation. Thus each part of the organism exhibits its peculiar phase of the time-keeping faculty, showing that time is a component of man as a whole. In proof of the separateness of Time as distinct from Tune, let me state that I have known those who were so destitute of Music as not to be able to know one one musical air from another, yet who kept time perfectly in dancing and marching. These were persons in whom the Bony and Muscular systems were in the ascendant.

As I have shown that Time inheres in the entire organism of man, it necessarily follows that there would be many signs by which to discern this faculty and to discover the particular sort of Time with which the individual is endowed. In accordance with the principles upon which it is founded, we should, then, find the kind of time used by musicians in the round form, and also in those who combine the Bony with the Muscular system, being so blended as to avoid both angularity and obesity. In those who apply time in mathematical computation, we shall find the square bony form in excess in some, and the Brain system predominant in others. Among musicians and composers of music there are many of the round build, like Handel and Meyerbeer; many with the Brain system dominant, such as Wagner, Von Weber, and Mendelssohn; others in whom the Bony and Muscular systems seem about evenly developed, as we find in the physiognomies of Listz and Haydn. One local sign for Time is found in the forehead, just above the sign for Order. This sign is discovered mostly in the muscular and bony people, and is owing to the peculiar conformation caused by the combination of these two systems. Dr. J. Simms gives as indications of Time the following: "Mechanical Time is known by a squareness of the face, joined with a large numerical capacity." Of the kind of Time denoted by the round form, he says: "The round form of the face and physique bespeak for the individual the ability to comprehend and produce natural time." His further illustration is as wonderful as it is beautiful and comprehensive, showing the law of correspondence with the shape and motions of the planetary bodies.

CAUSALITY.

SIR ISAAC NEWTON. (CAUSALITY.)

Newton possessed in a remarkable manner the power to trace cause to its origin. His large Conscientiousness was a valuable assistant to this faculty. To his Causality and love of natural truths we are indebted for the discovery of the law of gravitation, and other natural laws as well.

This is the faculty which traces causes to their origin, seeks out the why and wherefore, deduces inferences from premises, causes originality, and gives reasoning power and invention. Its principal sign is located in the upper and lateral portion of the forehead, on either side of Comparison, which is its natural ally in perfecting reasoning processes. Where Causality is largely developed, it produces a rounding out and slight fullness of this part of the forehead, giving this part of the face breadth. It is found larger in broad, strong organisms than in thin, weak persons. Its foundation is in general strength of the body and mind, with a suitable quality of the Nerve and Brain system. Its location is the highest in the face, and enables it to sit in judgment, as it were, upon all the acts of the body and processes of the mind. In conjunction with Comparison, it presides over the religious faculties; for "religion without reason is superstition." It is, in fact, the latest acquisition to the human mind, and the greatest distinguishing difference between Man and the brute creation; although there is no doubt that the animals of the higher grades reason in a limited degree. The facts of natural history attest this, and Man, with all his boasted superiority, is different only in degree; he remains the same in kind as other animals.

Causality is found in the faces of all who excel in investigation, research, reason, science, and jurisprudence. In the physiognomies of all the celebrated jurists of England—Lord North, Earl of Clarendon, Erskine, Blackstone, and Ellenborough—this faculty shines pre-eminent; also, in those of the most eminent scientists and astronomers—Lalande, La Place, the Herschels, Galileo, and Mitchell; among metaphysicians, in Hobbes, Paley, Adam Smith, Dugald Stewart, and John Stuart Mill; among divines, in Jonathan Edwards, John Knox, Melancthon, Wesley, and William E. Channing; among scientists and naturalists, in Ampere, De Candolle, Sir Humphrey Davy, Von Liebig, Buffon, and Agassiz. It is large, also, in the faces of the celebrated statesmen of all ages, such as Metternich, Talleyrand, Pitt, Fox, Palmerston, Webster, Monroe, Jefferson, Calhoun, Alexander Stephens, and others.

This faculty, along with that of its close companion, Comparison, is best illustrated where the forehead recedes slightly. This angle of inclination always discloses the practical reasoner. The foreheads of all great intellects indicate this formation. See, for example, Locke, Cicero, Newton, Cuvier; while the foreheads of many artistic persons are thrown more forward at the upper part. They have not the necessity for practicality as the former.

COMPARISON.

See portraits of Lord Chancellor Somers, Sir Isaac Newton, and Charles Bradlaugh, on pages 75, 117, and 127.

Comparison is the companion, naturally and necessarily, of Causality. Possessing high powers, it holds its position in accordance with its importance, being located between the two local signs of Causality, and above Memory of Events. It fills out and elevates the centre of the forehead at the highest part, and is most effective in those who are possessed of strong, broad, and harmoniously organized bodies, together with a high quality of nerve and brain. It enables the individual to criticise, analyze, compare, classify, and discriminate between truth and error. It endows him with the power to generalize and apply facts. It assists the scientist, philosopher, architect, and mathematician. All our great judges possess the faculties of Comparison and Causality in a pre-eminent degree.

These faculties are never seen with a flat nose, and all races and persons who have undeveloped noses are destitute of the gift of inductive reasoning, are never philosophical, and are not capable of performing the highest mathematical demonstration; for "the size of the nose is the measure of mental power," as well as the gauge by which we measure the architectural or building powers and artistic capacities in man. The portraits of Socrates, Aristotle, Bacon, Newton, Kepler, Voltaire, D'Alembert, Descartes, Condorcet, Hugh Miller, David Hume, Herbert Spencer, and Owen Jones show Comparison to have been very large in their organizations. All of these men possessed large, high, broad noses, and broad, full foreheads at the superior part. Causality and Comparison can always be inferred if the nose is high and broad at the bridge and below it.

INTUITION.

CHARLOTTE BRONTE.   (Intuition.)

Charlotte Bronte was distinguished for her insight into human nature.   Her analyses of character are remarkable when we consider the limited field she possessed for investigation.   She was an English novelist, one of the most eminent of the age.

Intuition is a faculty which seems to be directly related to the Brain and Nerve system, and depends upon a peculiar quality of this system for its power. If Intuition is found with persons of the Muscular, Thoracic, or Bony systems predominant, then its action is determined by the quality of the brain and nerves in conjunction with it. This quality manifests itself by keenness of apprehension of the character of everything of which the individual takes cognizance. Persons with this faculty large are good physiognomists; understand at first glance, almost, the physical conditions of those whom they observe. Such scarcely need to be told the symptoms of disease, but appear to apprehend the various feelings and forms of suffering, and to make a correct diagnosis of diseased and healthful conditions with facility. It is not adapted alone to the requirements of physical insight, but brings with it intellectual intuition, and comprehends instantly the indications and appearances of intellectual truth or falsity; it penetrates moral conditions, and is able to almost divine the true, and to separate it from the false system of ethics.

I believe the higher animals to be possessed of this faculty in a large degree. It is their method of discerning the character of their fellows, as well as that of their masters and human companions and acquaintances. This endowment is for their protection, the same as with Man. It can be cultivated by those who possess it, but cannot be imparted to others. Persons of the Vegetative system large have very little of this faculty; it is found mostly with the higher systems. The great actors depend largely upon Intuition to aid them in comprehending the characters which they personate.

The signs of this faculty are various. The principal facial sign is a bright, clear eye of a decided color, such as true blue or pure black or brown. Mixed eyes or very pale ones are seldom found with those whose intuitions are correct. Another sign is a predominance of the Brain system over all others. Still another sign is shown by fine, clear skin and fine hair. Suspicious and jealous persons have so little of Intuition that they substitute suspicion for observation and truth, and are jealous because they cannot apprehend traits as they really exist in the characters of those about them. Persons with large Intuition should study the natural sciences, because they are so well adapted to discern universal truth. They should devote themselves to scientific Physiognomy, because their natural qualifications will enable them to become experts in character reading, and the requirements for a first-class physiognomist are very exacting; no other faculty can be substituted for Intuition.

The face has for its expression thirty-six pairs and two single muscles. I advise my readers to procure some good work on Anatomy and Physiology, and make themselves familiar, not only with the muscles of the face and body, but also with the several systems included in the human organism; viz., the Vegetative, the Thoracic, the Muscular, the Osseous, and the Brain and Nerve systems. The scope of this work will not permit an extended anatomical and physiological description of the numerous principles and laws regulating and giving expression to the physiognomy. As in all sciences there are many points which can be imparted only by a teacher, so in Physiognomy there are many meanings to expression which cannot be reproduced on paper, but must be taught from the living subject. Still, enough of the science is given in these pages to lift the veil which has so long hidden Man from his own knowledge, and every careful observer and thinker will be able, with the assistance here rendered, to progress with rapid strides toward a more complete knowledge of this wonderful and beautiful science.

The faculties treated of in the Mathematical division are the highest in the scale of human development. I mean by this that this part of the face is developed only in the most perfected races and people, and appears latest in evolution. It requires a more perfected mind to be mathematical, in its highest and broadest sense, than to be either artistic or literary. A man may be highly religious, in the common acceptation of the term, without having a single idea as to the powers of numbers, or an appreciation of the harmonies of rhythm or chromatics, or of the grand and far reaching knowledge of Astronomy. He may be artistic to a high degree, without being able to comprehend the harmonies of Nature which bring to us a knowledge of the wonders of infinite causation. Causality and Comparison sit in judgment upon all the processes of mind and acts of body; their locality is above all the other faculties in the face. They are the only faculties through which we can comprehend the unity and co-ordination of the laws of matter. They are pre-eminently large in philosophers and in those who have given to the world systems of Causal and Formal Astronomy and Science.

Aristotle, Galileo, Bacon, Lavoisier, Cuvier, Linnæus, Kepler, and Newton must have presented, in their own organizations, such combinations of form and faculties as were in harmony with, and based upon, just such principles as the powers are which they discovered and demonstrated. Such, indeed, is the fact, as the "counterfeit presentments" of their faces and physiques show.

CHAPTER V.

EXPOSITION OF SUB-BASIC PRINCIPLES.

FORM AND SIZE.

Whether we accept the doctrine of evolution or not, we must, with the vast array of evidence in organized life before us, admit that there is a singular unity of action influencing the methods of Nature. An ordinary observer will find that certain characteristics in the animal outwork in the like results when found in the human family. The same general laws as to form, size, color, texture, proportion, and faculties are common to both man and the brute creation. A study of the various conformations, colors, and textures of the several species of animals, both wild and domestic, together with the birds—which, Geology shows, preceded the animal kingdom in the scheme of creation—will assist very materially in the knowledge and proofs of scientific Physiognomy.

First, as to FORM and SIZE. Consider the hippopotamus— bulky, unwieldy, slow, with large abdomen, small brain, thick hide, wanting in sensitiveness, useless for any practical purpose. Behold the elephant, a little in advance of the former; he, also, is built on the broad plan, with dark, tough skin; brain considerably larger, in proportion to his size, than that of the hippopotamus; much more intelligent, sagacious, and sensitive; bony structure more prominent, with dark eyes and skin; possessed of greater activity of emotions, revengeful under insult or injury; very wide mouth and large abdomen. To which formation of the human family do these animals correspond? If you have given attention to the preceding pages, you will recognize at once all the general characteristics of the Vegetative form; here is correspondence number one.

Examine closely the stag, made for mountain climbing; behold his length, leanness, activity, and form, the brightness of his eye, his ambition, desire for scaling the greatest heights, and his great breadth of chest, the broadest part of his body. He is here, there, and everywhere in a moment; does not dwell long at one place or pursuit. His lungs and heart must be well developed to give the power for such activity. This form is the counterpart of the Thoracic in the human organism.

Let us pursue this system of Physiognomy still further. In the animal world, whenever we see creatures endowed with the disposition for great destruction, we naturally look for a corresponding amount of strength; in this grade of development, strength and destruction are synonymous. If you were to examine a lion, tiger, or panther, you would find them characterized by strong, compact muscles, dark hairy coat, dark or yellow eyes, with rapid motions, intense passions, and great courage. This class of animals represents the Muscular system in man. Persons of this form exhibit great strength, capacity for destruction, and large amativeness. They are also social, domestic, emotional, and commercial; the commercial faculty corresponding to the preying and getting instinct in the animals of the same form.

As I have previously shown that the most reliable, moral, tractable, and naturally intelligent of the human species are found where the Bony system is predominant, so in the animal kingdom you will find the corresponding faculties in those domestic animals—the horse, cow, ox, camel, and dog—who render to mankind faithful, gentle, and intelligent service. The distinguishing marks as to color, form, and texture are relatively the same as in the Bony system predominant in man. The prominent points are rather fine hair, variety of colors, given to herbiverous living; although the dog, like man, lives on a mixed diet. The horse and dog are particularly receptive; the projecting bones over the eyes resemble the development of the practical or mechanical faculties in man. Width between the eyes, in either dog, horse, or man, is always indicative of a broad intelligence. It shows the faculty of Form to be large, and also gives breadth to *all* the functions and faculties of the mind; for Physiognomy, well understood, reads the body as well as the face; it takes cognizance of the color of the hair, skin, and eyes; it observes the walk, the voice, gestures, and movements. All are indices of character. To a practiced ear, the intonation of a single sentence will reveal very much to the listener. Everything which one does, no matter how trifling, is highly significant of character; and habits of observation and analysis should be formed in youth, and the reason why traits are combined as we find them should be given by parents and teachers. I feel assured that, after a careful reading of these thoughts, any parent will be competent to direct aright the dawning perceptions of his child in Physiognomy. It is the *duty* of all parents to throw around their children such protection as the knowledge of the laws of Nature affords. It will prevent the erroneous reading of character to which the present lamentable ignorance of the laws of Physiognomy leads. The many physiognomical errors current will be rectified, and the human family will be given a compass which will keep it clear of many shoals and quicksands which are found on the journey of life.

Size is a subject so little understood that I feel compelled to correct some opinions in regard to it, which have come to be accepted as truths. The phrenological principle that "Size is the measure of power (all else being equal)" has been accepted in part, and it is now generally thought that a large head is, or ought to be, the sign of a very intellectual person. Now, the truth is, that the larger the head the duller the person, *unless* there are very many favorable conditions accompanying it; and the first is inherited quality. Now, *quality* is the determining power all through Nature—not *size*. If one wishes a fine flower, one does not pluck a sunflower. It is large, true; but it is also coarse in look and devoid of fragrance. So one selects a smaller and more developed flower. This development is shown in the same way by which a brain or an ear of high quality is known—by the number of its convolutions. A fine rose or pink will illustrate this difference. A flat, broad ear, standing out at the side of the head like that of a pig, is no indication of "ability to receive fine musical sounds. On the contrary, a small ear, with a well defined rim and a number of elevations and depressions, will show itself to be of high quality." (Simms.)

A large head is a serious disadvantage to its possessor, unless it be accompanied by large lungs, large heart, and good digestive powers; also, inherited quality of a high order. Then, again, a great deal depends upon the shape or form of the head. If most of the head lie behind the ears, the character will be low and sensual; but if the top of the forehead be broad, and much of the size lie in front and above the ears, then the character will be pure and noble, as well as intellectual. A large head merely, without all or most of these conditions, is an indication of a dull, stupid person. A large head must be accompanied by large lungs, in order to supply a great deal of oxygenated blood. The brain requires one-fifth as much blood as the entire body, according to Haller, and it is apparent that unless there be an ample supply of this material, there can be but a low degree of activity in the brain. Peuschel, a German observer, says: "A forehead of excessive size announces a man slow of conception, dull or sluggish in forming his ideas." Schaliz, another observer, tells us that "a forehead too large is the sign of a character timid, indolent, and stupid." The physiognomical indication of greatness is not found in size alone, nor in any other *single* sign.

I have laid down, in the chapter on Basilar Principles, the rule that "the size of the nose, controlled by quality, is the measure of power; the shape of it is the proof of the kind of power." Therefore, in order to recognize powerful character in an individual, we must see that the nose stands high above the plane of the face, the nostrils broad, the eye large and bright, the mouth also large, the chin of proportionate breadth and length, the eyes set well under a rather project-ing brow (an eye that is on a level with the plane of the brow discloses great stupidity), the cheeks well filled—not fat, a forehead broad across its upper part; and when to this is added a fine skin and fine hair, true greatness of some sort is indicated. The kind of greatness depends upon the shape of the nose. If it be a literary nose, then the possessor will excel in a literary direction; if the nose be architectural, that power will be exhibited; an artistic or dramatic nose will decide the talent and power of the individual in that depart-ment. To make all this effective, good health is most im-portant, for without it the individual would be like a power-ful steam-engine without steam—an inert, helpless machine.

QUALITY.

In determining the quality or mental power of an individ-ual, the texture of the skin and hair is to be considered first, as these indicate quite as much as the form, and really de-termine its power and activity. If the skin be fine, clear, smooth, and thin, a high grade of mental activity may be inferred. As the brain-substance, in the form of nerves, is spread all over the surface of the skin, the thinner and finer it is, the greater is the amount of sensation experienced; and, as Nature is harmonious, all the external appearances will be found to harmonize; hence, the hair will agree with the skin in quality, as well as with the finger-nails. The latter will be found smooth, fine, and thin in combination

10

with a skin of like qualities. This way of deciding the *quality* of mental power is infallible. The peculiarities of the formation of the face must tell the rest. The same law obtains in the animal world; a fine, soft coat on any animal proves its superior intelligence to those who possess coarse, shaggy hair. The exterior will always be found to agree with the interior in quality and form; and, after we learn the indications, it will be astonishing how simple it will seem to read character correctly, and we shall wonder why we never saw these things before, nor fathomed their meanings.

The brightness of the eye is still another exponent of the quality of brain power. An eye that is dull naturally, and moving slowly, shows dullness and stupidity; while bright eyes, with a quick and animated motion, show that the mental powers are clear and active. There is much in regard to the eye which cannot be written. Words fail to describe adequately different degrees of brightness and expressions. The reader must investigate for himself, and commence a course of generalizing and classification on his own account.

I have spread before you, in the thoughts which immediately precede these, the similarity between certain forms, colors, textures, and faculties which are found both in men and animals. You will observe other principles active in all these organisms, if your attention be directed to them.

COMPENSATION.

All through the various forms of what may be termed the higher development of organized life—from the insect up to man—we find clearly established a law of Compensation, or, as I am sometimes impelled to call it, a law of Substitution; for its action does not seem always to *fully compensate* for absence of qualities, but rather *substitutes* other powers, both physical and mental, for defects which would render the organism helpless or unhappy without some assistance from other faculties and functions. In this relation I shall—as this is a very important branch of my subject—dwell briefly on its action in the lower organisms, and will then proceed to discuss its operation in the human mind and body. And here let me remark, that while the law of Compensation has been recognized by naturalists in the animal organism, it has never been applied scientifically to the workings of the human mind, so far as I have been able to learn.

The compensatory structure of animals will be easily recognized in the following statement of Paley. He remarks: "In many species of insects the eye is fixed, and consequently cannot turn the pupil to the object sought. This great defect is perfectly compensated by a mechanism not easily observed. The eye is a multiplying glass, with a lens looking in every direction, by which means—although the orb of the eye be stationary—the field of vision is as ample as that of other animals, and is commanded on every side. We are told that one thousand four hundred of these reticulations have been counted in the two eyes of a drone bee. The wing of a bat is furnished with a mechanical contrivance in the form of a hook, with which it fastens itself to the surface of rocks, houses, and caves. At the angle of the wing, there is a bent claw. It hooks and remains suspended by this claw; takes its flight from this position. As it can neither run upon its feet nor take its flight from the ground, this unique instrument was necessary. A singular defect required a singular substitute. The proboscis of an elephant is a compensation for the shortness of its neck. A snail is compensated by the secretion of a viscid humor which it discharges from its skin; and so, in the absence of feet, is enabled to ascend the stalks of plants with facility." I could multiply these examples *ad infinitum*.

In the human family the illustration of the law of Compensation is more extended, and includes the mental as well as the physical system. This involves some knowledge of the law of Proportion, or harmonious development of the body—upon which, of course, depends the harmonious action of the mind; for, as before stated, certain conformations of the body produce certain mental faculties. It therefore behooves us to know to which forms these faculties are related, and how produced.

I will notice, first, the operation of the law of Compensation as regards the human organism physically, or, rather, physiologically. In cases where one lung is weak, the other often increases in *size* and *power* to make up the deficiency. Deaf mutes are compensated by an increased activity of all the other senses. Blind people are unusually gifted with an acuteness of the senses of hearing and touch. Where the kidneys are small or weak, the skin is uncommonly active, and assists the kidneys in carrying off the waste of the body. These are some few of the ways in which Nature compensates for defective and inharmonious organizations. The manifold action of the law of Compensation as exhibited in the working of the human mind is as wonderful as it is beautiful. I shall have space here to offer only a few illustrations, and leave the rest for the investigations of my readers.

If you will observe an individual with very small Self-esteem, which is indicated by a short upper lip, you will find Imitation, and generally Mirthfulness, correspondingly large. The philosophy of this form of compensation is that as small Self-esteem produces sensitiveness to the opinions of others, Imitation seems given the individual to assist him in entertaining and attracting, while large Mirthfulness gives the faculty of amusing and being easily amused, and consequently prevents the individual from becoming unhappy through the consciousness of the absence of Self-esteem; for any deficiency which prevents a balanced condition produces a want which is *instinctively* felt. I hold that we all instinctively *feel what we are*, whether we acknowledge it to ourselves or not. Actors as a class possess the faculties of Imitation and Mirthfulness in a large degree, and most of them will be found deficient in Self-esteem, but large in Approbativeness; for it is not their own esteem that they desire and which satisfies them, but the approbation of their audiences. A large proportion of them have a short upper lip—relative length of the upper lip indicating a good development of Self-esteem. An individual with large Self-esteem being self-sufficient—that is, more given to regard his own opinion of himself that to accept the estimate of others in regard to his character—feels no particular sensitiveness as to what others think of him, and therefore depends upon himself, *just as he is*, for the power to attract and hold the esteem of others. Self-esteem lends dignity to its possessor, and induces a substantial and decorous demeanor, which, in itself, has the power to fasten the good opinion and attachment of others, and he therefore needs none of the fascinations of imitative talent to attract friends to himself. Indeed, every one could not be attracted by the same qualities, and so Nature gives this infinite variety and diversity for the satisfaction of our minds and for the varied uses of mankind.

Where Friendship is lacking, we often see Benevolence compensating the character. Where Constructiveness is wanting, Size and Form assist, by an increased development, in making the individual useful in some branch of mechanical art. This system of Compensation inheres in the entire mental constitution. Later on, I shall refer to this subject and to the localizing of traits. Enough, however, of the compensatory action of the mind has been shown to illustrate its method.

PROPORTION.

A correct knowledge of the laws of Proportion governing the human physiognomy and organism will not be found to accord with the laws of proportion as taught in the schools of art. Science has wrought a mighty change in nearly every department of knowledge. It is possible that a wide-spread understanding of the laws of physiognomy as manifested in Nature may also create a revolution in art. The Greek ideal of symmetry, to which the ages have given their assent, will be found to be based on mathematical calculation, and it is from this cold and mechanical idea of what constitutes beauty that the modern conceptions of beauty and proportion are taken. Winkleman, criticising the Greek profile, says, "The nearer the approach to the perpendicular, the less is there characteristic of the wise or graceful."

Philosophy was the first means of breaking down the barriers to a scientific understanding of the laws of matter, of mind, and of the relations of man to the universe. Philosophy is the vanguard, which, breaking down the preconceived erroneous ideas of beauty and symmetry, precedes the natural, and therefore scientific, understanding of what is real beauty and true proportion in man. Victor Cousin, from whose admirable "Essay on the Beautiful" I quote, says: "The most important element in the beautiful is the *moral idea*. Unity and variety should be impressed with it, and serve only to exhibit it. *The beautiful is simply moral beauty*."

A scientific comprehension of the law of Proportion as shown in the human face will unfold more beauties than Greek art ever conceived. My understanding of beauty, as disclosed by Physiognomy, is based on the idea that the moral beauty exhibited in the countenance and form constitutes *true beauty*.

True greatness in the moral, mechanical, and mental constitution of man is not accompanied by any such law of proportion as the Greek or any other school of art has set forth. Nothing is more indicative of selfish will and heartless character than the so-called Greek profile. Lavater, the great intuitional physiognomist, says, in discussing its signification: "Depraved is the taste which can call this graceful, and, therefore, it must be far from majestic. I should wish neither a wife, mother, sister, friend, relation, nor goddess to possess a countenance so cold, insipid, affected, stony, unimpassioned, or so perfectly a statue."

A scientific interpretation of the face will reveal more beauties than the ordinary observer has any idea of; for when he comes to attach meanings to expressions which indicate beauties of character, he will regard them quite differently than when in his ignorance they signified nothing to him; and when an intelligent observer looks with the eye of understanding upon the countenances about him, his sense of the beautiful will be gratified beyond expression. A new world will open to him, and I predict that with a general diffusion of physiognomical knowledge a complete revolution in religion, art, hygiene, and government will be brought about.

Proportion is as potent a factor in determining character as Form, Size, or Quality; and yet an arbitrary system based on mathematical measurement cannot be set up, for the reason that such great diversity of form and size exists in which *symmetrical character* is exhibited. If we were to form a standard of beauty, and take for the standard those faces in which the most moral goodness or power for usefulness was disclosed, we should then have a more elevating and intelligent model than those already observed, which teach that beauty consists in mathematical proportions mainly, and not in those proportions and expressions which reveal something inward, something of moral grandeur or useful talent of a high order.

As has been shown, each of the five systems of the body produces a form peculiar to itself, and every human being possesses an admixture of some of each of these forms. It will, therefore, be apparent to the observer that the law of Compensation is more potent in forming Proportion than any other factor. If these systems were always blended in every form in exact proportions, we might then be able to realize the ideals of art in living forms, but this would not produce that differentiation of types which is needed to supply the varied wants of humanity. To carry out the idea of "diversity in unity," which is the ruling idea observed in progressive Nature, we must have constant modifications, which will, of course, produce ever varying forms and countenances. This comprehensive differentiation results in higher development of species. It is a law throughout Nature that the greater the variety, the higher the power for development and progress.

The law of scientific proportion and beauty to be observed in the human face is illustrated in those countenances in which all of the features, working together, express to the scientific reader a balanced condition of the mind, and consequently of the body. In a subsequent chapter you will learn somewhat of the practical effect of proportion when I come to treat of the form, size, and locality of the signs of character. Well proportioned forms have excellent assurance of longevity, because the several parts of the body and mind being equal, the wear and tear of life, or any unusual strain upon the organism, will meet with equal resistance, and thus no part will give way first.

HARMONY.

The law of Harmony, as exemplified in the action of the human organism, is the last grand proof of that intelligence and design which preside over all of Nature's works. Every part of each separate individual is adapted to every other part of that individual. The physical harmonizes with the mental, and *vice versa*. When Nature creates great love of the artistic she provides some way for its expression, either by tongue, pen, brush, or chisel. Where she gives large capacity for mechanical invention, the physical frame will be found built on the plan which will facilitate the carrying out of this power to the highest degree. Whenever a man possesses large mental power for planning grand enterprises in commerce or war, his physical endowment is such as to assist in forwarding his schemes. And thus, all through the human organism this law of Harmony will be observed asserting its supremacy.

In the same way the external is always indicative of the internal. If you observe an individual in whom the nostrils are large, you will always find the lungs corresponding in size; if the nostrils are round, the lungs will have the same shape; if long and narrow, the lungs will possess the same form. In the same way, long, thin, bony fingers belong to a physique tall and bony. The science of Palmistry, once so popular, was based on this law of Harmony, and involved a knowledge of Physiognomy and Comparative Anatomy. When you come to understand the principles of Physiognomy, you will be able to tell much of the character from the hand, the foot, or from the hair, for the principles of Harmony are at the foundation, and the signs are infallible, the same as in all of Nature's works. You will never find coarse hair or coarse skin accompanying one possessed of great mental endowment. Whenever you discover the texture of the skin to be fine, you will find in the individual fine hair and decided mental activity and capacity. In this, as in all things, the external harmonizes with the internal, and is evidential of it. The proofs and indications of Harmony are all over the entire organism, as much in the hands as in the face, as much in the foot as in the figure. Indeed, a skillful observer, after careful study of these principles, and after due observation, should be able to tell from one finger, or a lock of hair even, very much of the individual to whom they belong. It is simply Comparative Anatomy put into practice.

COLOR.

Deep colors are indicative of heat, and therefore of great activity. All through Nature this principle holds good: that dark-skinned persons, with dark eyes and hair, have more intense passions and emotions than light persons. Love, jealousy, and revenge are all most active with dark people. It is the same with dark or black animals; a black horse is more fiery in his disposition than a white one, and less teachable. You will always see white or cream-colored horses employed in a circus as trick horses, on account of their superior intelligence and docility. Light persons and races are found to be more progressive than those of dark color. As their passions and emotions are not so intense, they are more capable of improvement.

The colors of the human race proceed from the minerals

in the earth, conveyed through the medium of foods, and from the air and light. Every mineral has its own color; the deeper the color, the more powerful its effect; just as with men of dark complexion, they have more powerful passions. The same law follows right up from the mineral to man. Color, like all the primitive characteristics of man, was intended, in the first place, as all evidence shows, for the protection and preservation of the organism. Those in whom the coloring pigment or matter is wanting are weaker than those who have a normal supply. We see this quite often in young persons who are growing too fast. Enough coloring matter is not taken into the system by the medium of the food or by exercise in sunlight; hence, the skin fails to get its proper supply. The pallor produced indicates enfeebled conditions of other parts of the organism.

Haeckel, in his "History of Creation," treating of the influence of color, says: "Very frequently, albinoes are more feebly developed, and consequently the whole structure of the body is more delicate and weak than in colored animals of the same species. The organs of the senses and nervous system are in like manner curiously affected when there is a deficiency of coloring pigment. The want of the usual coloring matter goes hand in hand with certain changes of the formation of other parts—for example, of the Muscular and Osseous systems—consequently, of organic systems which are not at all ultimately connected with the system of the outer skin." He also says: "White cats with blue eyes are nearly always deaf. White horses are distinguished from colored horses by their liability to form sarcomatous tumors. In man, also, the degree of development of pigment in the outer skin greatly influences the susceptibility of the organism for certain diseases; so that, for instance, Europeans with a dark complexion and brown eyes become more easily acclimatized to tropical countries, and are less subject to the diseases there prevalent—inflammation of the liver, yellow fever, etc.—than Europeans of white complexion, fair hair, and blue eyes."

All bright colored insects, beetles, butterflies, and birds have the color sense very large. They seek out the brightest flowers and foliage for their enjoyment, and thus this sense is increased and transmitted from generation to generation. Tropical forests, which they love to frequent, must be gorgeous in the extreme, with their brilliant flowers and foliage and their many beautifully colored birds and insects. To one with the color sense largely developed, this sight must be inexpressibly beautiful and satisfying.

The secondary use of color in man is to assist him in comprehending its nature and harmonies for industrial and artistic purposes. Deep color in the skin, hair, and eyes is indicative of a love of color and an appreciation of its harmonies and blendings. I venture to affirm that no great color-artist ever existed who was possessed of very fair hair and very light eyes and a colorless skin. Lavater understood this principle, for he says: "We read the coloring of Guido and Guericino in their countenances." Dr. Simms also lays this down as a principle in man's organism, and gives as a sign for the color sense, "deep color of the hair, eyes, and complexion." My experience has led me to observe that the color sense may be imperfect where the eyes and hair are dark, and the skin pallid or not clear. It is necessary that color should be well defined in the entire organism to give the color sense its highest power. Persons who have the color sense the best developed are, without doubt, those who have inherited large, strong lungs. This enables them to inhale copious draughts of air, which serve to oxygenate and thus color the blood. By this process, the color of the skin and eyes is deepened, and thus the color sense is enhanced.

The creatures which are best endowed with the color sense, next to man, are the birds. Their existence is passed in the open air; the atmosphere permeates their entire body; even the bones of birds contain air-cells. These, undoubtedly, serve a two-fold purpose: they assist the bird in flying, just as the lungs assist man in swimming; and they also serve to carry oxygen into its organism, and thus assist the lungs in their work of oxygenation. The air thus introduced conduces to color their plumage, and this being transmitted and intensified by inheritance, it is not strange that the birds whose lives are passed in the higher atmospheres should be the finest colored. Marsh birds, and those who live near the ground, are not so highly colored as those who live upon trees and in the higher atmosphere. Flowers, likewise, being always exposed to the sun's rays and to the open air, are gorgeous with the hues of the rainbow.

This exposition of the sub-basic principles of Physiognomy will teach the reader that, in the task of analyzing character, very many principles are involved and must be considered in order to render a just reading of the face. Besides the requisites here mentioned for consideration, there are many expressions which have been acquired by long use or misuse, which always leave their impress indelibly stamped upon the countenance. A man can no more work as a blacksmith for years without showing the increase of muscle in his arms, than can · one use constantly the same set of muscles in the face without their leaving a permanent indication of such use. By watching closely the movements of the mouth in talking, one can form a very good estimate of the kind of language which that mouth has been accustomed to utter— whether it be kind, gentle, and loving, or cross, peevish, bad-tempered, and profane. The record is indelible, and cannot easily be erased or changed, except by long practice in another direction. All abuses of the physical functions write their record upon the face. The dram drinker, the sensualist, the glutton, as well as the sneak and liar, may all be detected by a close observer who has learned to apply the rules of scientific Physiognomy.

Of this tendency of the muscles to reveal long continued states of mental and physical abuse, Dr. John Cross remarks: "It lies with Physiognomy to detect the impostor; for, however well he may manage to jabber about morality, honor, or even religion, yet he cannot hinder the muscles without from obeying the central impulse; nor can he prevent an organ whose function is perverted from falling, according to the self-accommodating power, into color, size, and shape most suitable to the performance of this perverted function."

CHAPTER VI.

THEORIES OF CERTAIN TRAITS.

" No impartial judge can doubt that the roots, as it were, of those great faculties which confer on Man his immeasurable superiority above all other animate things are traceable far down in the animal world."—HUXLEY.

This age is peculiarly one of invention, of scientific research, investigation, and demonstration. The invention of the numerous and varied instruments used in the discovery of the laws and application of the apparently inexhaustible forces of Nature proves to us that there is nothing created in vain. Recent discoveries in light, color, sound, and the atmospheres are opening to us a world composed of the most subtle powers in the great laboratory of Nature. Examine them as we will, destructive as many seem, they have each a use in the great scheme of Nature. Electricity is a creator and a destroyer; light tears down and rebuilds; the atmospheres tend to both life and death. The forces which seem beneficent act also a malevolent part. Why is this?—why does God permit sin?

These are questions which theologians have grappled with, unsuccessfully, for centuries. It is only the scientist who, aided by a persistent and intelligent "interrogation of Nature," can answer these questions. The invariable conclusion will be, that everything has its use and place in the world; that nothing is made in vain; that thunder and lightning are useful; that birds and beasts of prey are necessary. Even snakes, gnats, flies, fleas, and other destructive and annoying creatures, have their use in the world.

So, in the human family, all those passions which, unrestrained and not balanced by justice and reason, cause destruction and suffering, are just as necessary and just as useful in the human economy as the most moral and intellectual traits. Jealousy, revenge, suspicion, resistance, secretiveness, and conceit are, in the present degree of human development, a natural and necessary accompaniment. We are in the transition state, moving from the lower to the higher. Human nature, like all growths, has its order of progress, marked by laws which are unerring. It is our province to seek these laws and apply them, in order to facilitate man's rise to that high and holy estate which is his destiny.

The first step toward this much desired result must be to understand the meanings of the forms and faces about us; next, what causes produce them; and then, to make use of this knowledge to create higher types. The only reparation we can make to the world for our failings, is to assist in perpetuating a race which shall be as noble as the laws of science can create. How many persons, observing the action of love, jealousy, revenge, suspicion, secretiveness, self-conceit, and the like, stop to reflect, for one moment, on the cause or *rationale* of any of these passions and traits? The major part of the world live in their instincts, as do the animals, but without the restraints which hold the animal to the due observance of the laws of his being, and which prevent him from making the stupid and miserable failures in modes of living, propagation, etc., which man, with all his boasted reason and freedom of action, is constantly repeating over and over again. Most persons love and perpetuate the race *instinctively*, without seeking or wanting any other guide than their feelings in the matter. Is this worthy such an exalted character as the latest development of Evolution claims for himself?

LOVE, or AMATIVENESS, is the fundamental faculty of the organism. Like all other faculties, it has its physical and mental aspects. In its normal development, it is the most beautiful and conservative of all the traits. It binds together hearts and homes, which serve to make the foundations of society and government sure. Like all other faculties, it is exhibited in different degrees and manner by each individual. The location in the face is in the chemical or moral group, and in close proximity to Love of Children, Mirthfulness, and other domestic faculties.

When possessed in a large degree, in combination with Constructiveness, it is most potent in producing the varied kinds of creative talent and art; and all who have excelled in the originating of ideas in every department of literature, in sculpture, in painting, and in dramatic representation or fiction—in short, all those who have shown themselves *creative* to any great degree—have possessed the procreative power in their physical organization in a marked manner. Exhibited largely, and with a moral balance, it makes the man very much of a man, the woman very much of a woman; and such persons will be more influential in their community than those deficient in this faculty. The latter are the small and impoverished characters one meets with, each hating the opposite sex, being hated in return; and this arises from the fact that such are not sufficiently sexed to appreciate their opposites. This faculty, exhibited in its physical development, without a balancing degree of Conscientiousness, leads to licentiousness and a violation of Nature's laws, and these are sure to entail suffering on its possessor and on all who come under its influence. This should warn us to observe the law of Nature in regard to the normal use of this faculty, for every function has a law for its government and protection. Each should seek this law for himself, since that law which may be binding on one does not necessarily involve every organization; although the Seventh Commandment should be binding on all. Each has a law peculiar to his own organization, which should be religiously observed. Indeed, religion should commence with the perpetuation of the race. I refer now to that religion which is the living up to natural law, and which, if rightly understood and observed as the laws of Physiology and Hygiene teach, would soon give us a race born under the law of true religion, who would become a blessing to themselves and the world at large.

JEALOUSY is commonly thought to be the necessary accompaniment of Love, and a proof of it. A scientific analysis of this passion shows it to be the result of an unbalanced condition. Wherever we find perception or justice lacking, or where the reasoning powers are not active, we will find this trait running riot. Shakspeare says of this passion, that "it makes the food it feeds on," and this is usually the case. "Perfect love casteth out fear," and perfect love has no need of such base company as jealousy. It is a standing insult and menace for one individual to be constantly jealous of another without exact proof. Persons with small Self-esteem are also subject to this passion. They are so constantly depreciating themselves that they naturally and instinctively infer that *any one* else is preferred to them; that is, they *feel* it to be so. Of course, they do not reason on it, not knowing the philosophy of this trait, and not knowing, either, where to locate it in the face.

REVENGE, like its kindred passion, Jealousy, is more largely developed among the dark races than among the light people; for, as in the animal kingdom, the darker the skin the less developed the organization. So, also, is jealousy more active when found among dark-skinned people, with dark or black eyes. I have never seen this trait in excess in a well balanced organization. We will often find it large in those whose will is in excess of their reason and justice. Their "will is law" to them, and when they cannot enforce it upon others they seek to be revenged, believing that they are wronged. In some, a deficiency of the practical faculties will cause this trait. This defect prevents the possessor from seeing the acts of others in their true light, and he consequently thinks himself an injured individual, and meditates revenge for his supposed injury. This trait is found most active with muscular people, especially if they be dark, and is often accompanied with a large degree of Secretiveness.

Whenever SECRETIVENESS is observed in an unusual degree in an organization we naturally infer that there is something to conceal, something deficient for which secretiveness is the compensation. It is a fine veil which Nature gives to hide a defect in either the mental, moral, or practical part of the organism. Some beasts of prey possess this faculty in a large degree. This is their normal condition. Having no mental or mechanical powers, as has man, to assist in procuring food, this faculty is needed by them for this purpose. Lions, tigers, wolves, cats, foxes, opossums, and all animals with the muscular system predominating, are most largely endowed with this propensity. Like its kindred passions, jealousy, revenge, and suspicion, it proceeds from a want of balance in the faculties; a lack of proper development of the reasoning faculties, Causality and Comparison, will produce it; a deficiency in Friendship or Human Nature will cause it; but wherever it is manifested one or more of these deficiencies will be found. Want of common honesty and uprightness of intention is sometimes the reason that Nature has provided this veil to assist the unfortunate possessor in making his way through the world. Secretiveness is given to animals to enable them both to avoid and to prey upon each other. Many persons having this trait are often considered very wise, owing to the careful and deliberate manner which they use in conversation. It is well that Nature has put this check upon their tongues, for if reason, justice, perception, or friendliness do not accompany the utterance of their thoughts, they would inevitably be led into more trouble than they could easily extricate themselves from; hence this check. Some mistake cunning or craft for wisdom. With persons in whom Secretiveness predominates the flexor muscles are more active than with others, often inducing a constricted state of the bowels and glandular sys-

11

tem, and particularly affecting the liver, causing biliousness, jaundice, and other derangements of this organ.

SUSPICION.—Comparison and Causality working together will arrive at and comprehend motives. The reason why one suspects the action of others is because he does not possess sufficient Causality to perceive the cause of their actions and motives, and therefore substitutes his suspicions; or he may be possessed of so little sense of justice as not to be able to comprehend common honesty in others, and therefore suspects that they are like himself; or else the perceptive powers may be lacking. But whatever produces suspicion, a defect will always be found in the organization as the exciting cause. In a work of this size it will be impossible to state at length these theories. Enough is given to show the methods of the action of the various passions in the human mind and body, and to arouse the student to the observation and research necessary to a correct understanding of these traits.

ANGER, WILL, or TEMPER, as it is sometimes termed, is a hydra-headed monster, manifold in its motives and action. Most phases of anger are detrimental to mental power and destructive to health. Only what may be called "righteous indignation"—that is to say, the indignation resulting from perceiving an infraction of the laws of justice or morality— is ennobling to the individual and conduces to strengthen both health and moral perception. This is the legitimate use of anger, and it should be reserved for such purposes. To become enraged at animals is at once wicked and stupid, and serves to show the superiority of animals to man. Nothing indicates the coward more than cruelty to our domestic animals, who give us faithful, gentle, uncomplaining service, and often die in harness while working for our benefit. The law justly takes cognizance of such treatment. These creatures are of our own flesh and blood, and we are not their equals in some things, although we may possess some qualities which are superior, but treating them cruelly and inhumanly is not the way to prove it.

SELFISHNESS is one of the traits of human nature which has two entirely distinct and opposite methods of action and purpose—one of which may be commended, the other reprehended. Selfishness, like all other faculties, has its use and purpose in the human economy. Its primal and essential property is the preservation of the body, and to provide for its perpetuation and maintenance. Its next legitimate use is for the protection and sustentation of those who are dependent upon us. All manifestations of selfishness which seek to please self, and to acquire by the suffering, misery, and unhappiness of others, are wrong and should be repressed. Speaking for myself, if I wished to pursue the most selfish course with the view of gaining the most, I would act the most unselfish and benevolent part in order to gain my purpose, for we get in this world very much what we give. If we strew our pathway through life with love, kindness, sympathy, noble deeds, justice, and gentleness, we will receive back the same with interest; but if, on the contrary, we pursue a malevolent career, and deal out hatred, malice, contempt, jealousy, suspicion, secretiveness, and anger, we will reap a harvest of these passions a thousand fold.

An undue degree of selfishness is indicative of an undeveloped nature. This trait is both inherited and acquired; increases by use, and in excess causes unhappiness to its possessor. The most selfish people are never the happiest; they cut themselves off from the pleasures and enjoyments of the benevolent, and thus limit the range of their happiness. They belong to that class which Lavater describes thus: "Which desires much but enjoys little, and whoever enjoys little gives little."

I have never studied a character which possessed an excess of selfishness that did not have also some serious deficiency in the mental or moral construction. Like the other passions treated of in this chapter, it shows undevelopment. The dark races are, as a rule, more selfish than the light ones; dark people are more selfish than light ones. They are less perfect, less progressive, generally.

All Nature attests this truth, that the more refined the person the lighter the color; It is the same with animals. The most destructive, revengeful, and jealous are the darkest, while the white or mixed colors are the most docile, amiable, and teachable. This is a general principle. Of course, there are exceptions; some undeveloped light persons being more selfish than very highly organized dark persons.

The excessive exercise and indulgence of jealousy, suspicion, secretiveness, and anger produce morbid and abnormal conditions of health, and herein is another proof of the relation of the physical organs to mental conditions. Many infants, even, have been made ill with jealousy by the petting and attentions bestowed by the mothers or nurses upon another child. Anger indulged in has wrecked the health of many. Suspicion turns often to insanity, and secretiveness to nonentity almost. Jealousy, the meanest and lowest of the passions, leads to murder and suicide, and self-conceit in excess to insanity. These excesses should be avoided, not only for our own preservation, but for the sake of those who are to inherit our individuality. All traits that are cultivated and indulged in are transmitted with increased power, and we have in this way the ability to become the benefactors of the race or to curse it beyond redemption.

Hippocrates, the celebrated Greek physician and physiognomist, says of envy: "The effects of envy are visible even in children; they become thin, and easily fall into consumption. Envy takes away the appetite and sleep, and causes feverish motion; it produces gloom, shortness of breath, impatience, restlessness, and a narrow chest." The possession of all these passions is antagonistic not only to the health of the possessor, but very much against his interest. Their action produces misery and unhappiness, both to the subject and the object. These conditions can be remedied by seeking out the defect, and making a constant struggle to correct it.

SELF-CONCEIT is perhaps the most harmless of this class of traits, but at the same time is ever offensive. Like all other faculties, it has its use and purpose. Nature has made nothing in vain, and so there would seem to be wisdom even in giving one an undue share of this petty trait. Where it is observed to predominate in an organization, it will be found to proceed from a lack of balance, as in the case of the preceding traits mentioned in this chapter. It is sometimes caused by merely a want of good taste, a deficiency in Ideality, by a lack of the perceptive or reflective power, or by dense obtuseness of the mental faculties generally. There are various other causes which produce it; but whatever the cause, it is designed to make up to its possessor the absence of *something*, which, if felt too keenly, would render him unhappy; so, Conceit, coming to his relief, puts him "on good terms with himself," and therefore has its use.

I have sometimes observed this trait very large in persons possessing real merit in some directions, but lacking in others. Conceit gives a sense of self-satisfaction, which is needed by its possessor just as long as he has the deficiency for which this is the compensation. If, on learning that he has a defect, and in what it consists, he would strive to remedy it by cultivating the defective trait or traits, he would soon be able to develop a more harmonious condition, and conceit would diminish or disappear entirely. Dwarfs and deformed persons are invariably conceited; the compensatory power of self-conceit in these cases is well illustrated; in such it is useful, and prevents unhappiness.

All these deficiencies can be remedied in a great degree, and sometimes eradicated, by a careful and scientific analysis of character and a settled determination to improve it. The laws of Physiognomy, thoroughly comprehended, will be the guide to that result; individual determination must do the rest. This improvement must be undertaken in a religious spirit, reflecting that all our actions, mentally, morally, and physically, affect not only ourselves, but go down to posterity, and curse or bless, for ages to come, all who inherit our blood even in the remotest degree.

The man whose life is passed with reference only to himself, without regard to children and children's children, is little better than the brute creature; in some respects he is worse, for the brute is not characterized by any such selfishness as this course would imply. There can be no motive more honorable in man than the desire to transmit to his offspring great and noble qualities, and this result can be obtained only by leading an honorable and noble life. We may endow offspring with fortune, but nobility and talent must be inherited; they cannot be bought in the market.

CHAPTER VII.

RATIONALE OF PHYSICAL FUNCTIONS AND THEIR SIGNS IN THE FACE.

"Whether the soul be air or fire, I know not; nor am I ashamed, as some men are, in cases where I am ignorant, to own that I am so."—CICERO.

"It will be understood by the word Mind we do not designate the intellectual operations only. But the word Mind has a broader, deeper signification; it includes all sensation, all volition, and all thought; it means the whole Psychical Life. And this psychical life has no one special centre; it belongs to the whole and animates the whole."—GEORGE HENRY LEWES.

The plan of this system of physiognomy would be incomplete were I to omit the *rationale*, or theory, of the action of the several organs and systems of functions comprised in the human body, and which assist in producing the various social, moral, and mental phenomena observed in the actions of the individual. Many philosophers have endeavored to ascertain the basis of mind; and by mind I mean that class of phenomena called reason, sentiment, mental operations, morality, the emotions, the passions, such as anger, jealousy, fear, hope, love, friendship, etc.

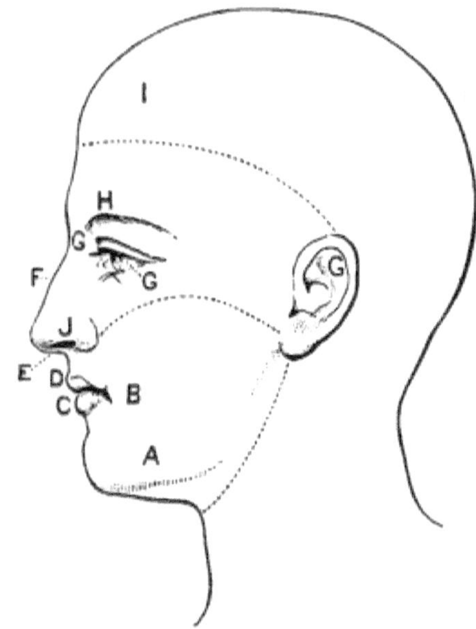

LOCATION OF THE SIGNS FOR THE DIFFERENT VISCERAL ORGANS AND BODILY FUNCTIONS.

A—The Kidneys. The sign for the kidneys is known by the width of the bony structure of the lower part of the chin.

B—Sign for Digestion, shown by fullness of cheeks at this point.

C—The Glands. A full rolling moist under lip indicates a good development of the glandular system.

D—The Reproductive System. Full red moist upper lip is its facial record.

E—Sign for the Liver. Where the septum of the nose projects well downward the liver is strong and active.

F—The Stomach. The higher and broader the nose at this point the more powerful is the stomach.

G—The Muscular System, shown by the size and fullness of the eye, the height of the nose between the eyes, and the thickness of the ears.

H—The Bony System. The projection of the bones of the brow indicates the dominance of the osseous system.

I—The Brain and Nerves. The more height and breadth in this division of the head, the more power for thought and sensation is exhibited.

J—The Lungs and Heart. Full wide nostrils, with good color of the skin, denote both lung and heart power.

The action of these is generally considered to be the result of brain or will power, with which the interior organs of the body have little or nothing to do. Theologians teach us that

the influences prompting many of the emotional states, such
as anger, hatred, revenge, jealousy, and the like, are created
by a spirit denominated a "devil." The acceptance of this
theory would end all further inquiry on the subject. My
observations do not corroborate their explanation of these
phenomena, and I am consequently forced to bring against
this view the Scotch verdict of "not proven." My theory of
the passions, so-called, will be found in the chapter on "Theories of Certain Traits," and the face read scientifically will
corroborate this theory.

The more recent of the philosophers and writers on the
origin of mind—Spencer, Lewes, Haeckel, Lindsay, and
others—have adopted the plan of seeking for the constituents and origin of mind by the investigation of matter; viz.,
in the *bodies* as well as *brains* of animal and human organisms.
And here I believe the problem will be solved. The introduction of words into our language, representing ideas which
are as far as scientific demonstration is concerned entirely
without foundation or support, has caused much confusion in
the minds of the masses of mankind. Before proceeding in
this study, the *idea* of "soul" as being in any way related to
mind (for at present we can offer no scientific proof that it is
anything but an idea) must be dismissed. This will•clear
away the hinderances, so that mind can be demonstrated
through the action of physical phenomena entirely, and without the complications and confusion which would ensue were
we to endeavor to prove the origin of the mind by mysterious doctrines dependent entirely on speculation and faith for
their explanation.

The brain has been considered by most metaphysicians,
philosophers, and anatomists even, to be the *sole* source and
seat of the mind. Recently a dim suspicion has been creeping into the minds of the more advanced and intelligent observers and thinkers that this may be an error. The proofs
of the theory of the soul and mind, so much dwelt upon by
the ancient metaphysicians, have no material or tangible
basis upon which to commence experiment and demonstra-

tion, but rest entirely on belief or faith. Hence, in the investigation of mind, we are necessarily limited to the observation of matter. By confining ourselves to this domain, we shall reach conclusions which I believe will be decisive.

The cerebrum, or front portion of the brain, has for a long time been considered by anatomists as the locality where thought, emotion, volition, and sensation are in some way (unknown) brought into a condition called consciousness. By recent experiments upon animals, and through accidents to human beings, it is demonstrated that the cerebrum does not possess the power formerly attributed to it. Much of it has been removed without destroying life and without causing the cessation of the principal physical functions. Indeed, in one instance, well authenticated by Longet, as quoted by Lewes, it is related that "a new-born infant, whose brain during the birth had been completely extirpated (to save the mother's life), was wrapped in a towel and placed in a corner of the room as a lifeless mass. While the surgeon was giving all his attention to the mother, he heard with horror a kind of murmur proceeding from the spot where the body had been placed. Soon a distinct cry was heard, and, to the surprise of all, this brainless infant was seen struggling, with rapid movements of its arms and legs. It cried and gave other signs of sensibility for several minutes."

Mr. Lewes also gives an account taken from the experiments of Bouillard, a French anatomist, who, he says, removed the whole of the cerebrum from the brain of a fowl, and he thus records his observations of this case:

"This fowl passes the greater part of her time asleep, but she awakes at intervals and spontaneously. When she goes to sleep she turns her head on one side, and buries it in the feathers of her wing. When she wakes she shakes herself, flaps her wings, and opens her eyes. In this respect, there is no difference observable between the mutilated and the perfect bird. She does not seem to be moved at all by the noises about her, but a very slight irritation of the skin suffices to awaken her instantaneously. When the irritation ceases, she relapses into sleep. When awake, she is often seen to cast rapid glances here and there. If put into a cage, she tries to escape, but she comes and goes without any purpose or rational design. When either wing or foot is pinched, she withdraws it. When she is laid hold of, she struggles to escape and screams; if severely irritated, she screams loudly. But it is not only to express pain that she uses her voice, for it is by no means rare to hear her cackle and cluck spontaneously; that is to say, when no external irritation affects her. Her stupidity is profound; she knows neither objects nor persons, and is completely divested of memory in this respect. Not only does she not know how to seek or take food, she does not even know how to swallow it when placed in her beak; it must be pushed to her throat. Nevertheless, her individuality, her movements, her agitation, attest that she feels the presence of a strange body. In this recital the evidence both of *sensation* and *instinct* is incontestible to any unprejudiced mind."*

Dr. Dalton, in giving the result of experiments he performed in removing the cerebrum of a fowl, says that "it was not accompanied with the loss of sight, of hearing, or of ordinary sensibility. All of these functions remained, as well as voluntary motion."

This is a mere allusion to the mass of evidence observed and collected by different anatomists, all going to prove that the brain is *not the exclusive seat* of sensation and consciousness. I advise my readers to consult the book from which these examples are taken, as well as the later work by the same author, entitled "The Physical Basis of Mind." I shall bring more of this writer's thoughts in support of the positions I take in this system of Physiognomy. My readers will pardon the extensive quotation I shall make, for I deem it only just to them, to my system and myself, that my theories should not lack competent authority, and that my ideas may not be accepted upon my unsustained observation and explanation alone. Many of them may seem novel; yet, upon

* G. H. Lewes's " Physiology of Common Life."

a close investigation of my premises, they will be found to derive their origin and support from Comparative Anatomy, Physiology, and cognate sciences.

Let us listen again to Mr. Lewes, whose opinions and deductions I value very highly. Comparative Anatomy, he says, "is freely invoked when it can sustain the argument in favor of the brain or cerebrum being the *sole seat* of intelligence, but it is quietly disregarded when it flatly contradicts the idea of the brain being the *exclusive seat* of consciousness or sensation. We cannot allow two weights and measures; if the evidence furnished by animals is good in one case, it is good in another. Now, what says evidence? A survey of the vertebrate classes discloses a remarkable correspondence between the size and development of the cerebrum, and the energy and variety of the mental manifestations. As we pass from fishes and reptiles to birds, and from birds to mammals—from the less intelligent to the more intelligent—we notice a decided increase in cerebral development. It is a legitimate inference that the one is in some correspondence with the other, and that intelligence is *one* of the functions of the cerebrum. Let this be admitted without reservation, although the well informed anatomist may have many difficulties to propound. I now ask what we are to make of the fact that multitudes of animals have no cerebrum at all, and that even among fishes there is at least one known to be without a vestige of it (and zoologists may discover many more); so that, unless we pronounce the amphioxus and all the invertebrates to be mere machines, without sensation or consciousness of any kind, we are forced to admit consciousness in the absence of the very organ which is said to be *its exclusive seat*. There are two answers open: First, it may be said, as it commonly is, that these animals have no intelligence—only instinct. This may be true; but to make it of the slighest use, we must be taught what instinct is, and that teaching is yet to seek. Instinct, like chance, is one of the words under which men seek to conceal their ignorance from themselves. That the actions of a bee or a crab, which manifest sensation, memory, invention of new methods under new circumstances—not to mention anger, desire, and some unexplained mode of communicating with each other—that these are actions of 'blind instinct' might not be inconceivable if we knew what instinct really was; but we shall find it difficult to conceive how precisely similar phenomena are attributed to intelligence when displayed by the dog or monkey. It is probable that the bee or the crab has no power of forming abstract propositions; it is probable that they are unable to carry on trains of thought remote from the sensations which are immediately affecting them. Let us, for a moment, grant that no invertebrate animal has intelligence, in any sense in which it has pleased man to employ the term. Let instinct explain everything, without itself needing explanation. It will not remove an iota of the objection against the assumption that the cerebrum is the *exclusive seat* of sensation and volition. The bee may have no vestige of intelligence, but you cannot deny that it has sensibility and volition; the brainless amphioxus may be a very stupid fish, indeed, but you will hardly assert that he wants the consciousness, the sensiblity of other fishes. If you grant me this, dispute is at an end; you merely say that the cerebrum has certain *special* functions, among which *intelligence is one;* you do not thereby exclude *from other ganglia* other forms of sensibility."

Mr. Lewes says, also: "In the same way, the brain of a bee is analogous to the brain of a bird. There are many and important differences, but there are fundamental resemblances of structure and connection of property and function. It is because they are both formed of ganglionic substances that they have both the property of sensibility. It is because they are both connected with the organs of special sense, and are the chief centres with which, directly or indirectly, all the nerves are connected, that they both have the supreme function of cerebration. This is the teaching of Comparative Anatomy, and its lesson is valuable if it shows us how the cephalic ganglia of an insect may represent the brain of a vertebrate animal, and thus *seem* to justify the doctrine of the brain being the exclusive seat of consciousness. It also, and by the same evidence, shows that sensibility must belong to all ganglia as ganglia, and not to any special group. The brain derives its sensibility from its ganglionic structure, in which it resembles all other ganglia; it derives its *functions from the various organs to which this sensibility is made subservient by anatomical connection.*"

Elsewhere, Mr. Lewes remarks: "I conceive, therefore, that Comparative Anatomy irresistibly disproves the notion of the brain, or any other ganglionic mass, being the *sole and exclusive* seat of sensibility or consciousness. No one will understand these remarks to mean that the brain is not one great centre of sensation and volition—the chief and dominant organ of the whole psychical mechanism. I have said before that it has the noblest functions, *but it does not exclude the other ganglia from their share* in the general consciousness. In it, all the sensations derived through the senses and viscera are summed up, combined, modified, and, in some profoundly mysterious manner, elaborated into ideas. In like manner, emotion may be considered as the form of cerebral sensibility *which is determined by connection with the ganglia of visceral sensation.*"

Let every fair-minded, unprejudiced person ask himself this question: For what are the ganglia connected with the several visceral organs?—what is their use? Why, says popular opinion, to carry to the brain the knowledge of the condition of those organs. Is that all their office?—is there no power evolved from these organs?—do they not sustain or create and nourish certain so-called "mental faculties"? Whence, then, is derived the sentiment of Love, for example?—is it manufactured in the brain, and exhibited only by the voice, by sentiment? If this were the case, then it would result in words only. This sentiment of Love is derived, in my opinion, from a physical base—from the functional action of the reproductive system—and results, in most cases, in functional activity of this system by reproduction. I think the most superficial reasoner will not dispute this. Now, if *sentiment* is derived in one instance from the *functional action of one visceral organ* and its ganglia, or plexus, would it not be corroborative evidence as to the *ability of all the other viscera to produce or create other kinds of sentiment*, such as Friendship, Conscientiousness, Love of Young, Benevolence, or Cheerfulness?—which last many of the most ignorant, even, understand is in some way connected with a healthy condition of the liver; for when they observe one who is "blue," as they express a despondent state of mind, they invariably ascribe it to a "bilious" condition of the liver, and correctly so; for Hope, which creates a happy disposition, is directly related to the liver; and if the sentiment of Hope depends upon the normal action of that organ, how can it be said that Hope is a *mental* attribute, and created in the brain? I grant that the liver must be connected with the brain, as we know it is by the great sympathetic or nervus vagus; but I deny that Hope is manufactured there. Its seat and source is in the *liver*, and depends upon, first, its natural construction, or size and quality; and, second, upon its normal condition. These two requisites being had, we find a cheerful, hopeful individual, with a clear, fertile, suggestive mind—so clear, indeed, as to make him highly analytical in everything which he observes or does. I know all this is antagonistic to the popular idea of mind, of sentiment and emotion; but whence, I ask again, does Mind derive its power? Not from the brain alone, because I have given you the evidence collected by such eminent students as Lewes, Dalton, Longet, and Boulliard, and the opinions of others as learned will follow this. I have shown that, in the case of the newly born child, movement, respiration, and vocal exercise were possible without any brain whatever. Now, if this be possible without brain, then the power was derived *from some other source*. I claim that it came from the several visceral structures; and the face, which is an exact register of the size and power of the various organs in the body, will prove to any good observer, who cares to investigate the science with a dispassionate mind, that where the signs for size of certain organs are found in the face, the mental characteristics which I claim are related to these organs will be found coexistent in every instance. Is this fancy or imagination, fact or fiction? The proof is within reach of every reader; let him justify my theories, or disprove them by evidence as conclusive.

Those who have passed years in the study and investigation of any branch of science are presumed to be more learned on the subject of their pursuit than those who have given it little attention, and I hold that the opinions of the former are entitled to the credence and respect of the latter. Believing this most fully, I append the following extract from the work of Dr. W. Lauder Lindsay, entitled "On Mind in the Lower Animals." It will not be without interest, and is entitled to our respect, in consideration of the source whence it emanates. Dr. Lindsay has been for many years at the head of an institution for the insane in Scotland, and is also a Fellow of the Royal Society of England. His investigations of diseased mental peculiarities of the insane have opened the way to an understanding of the *locale* of the mind, and he states his belief of its location and action thus. He remarks: "The student of Comparative Psychology cannot too soon divest himself of the erroneous popular idea that brain and mind are in a sense synonymous; that the brain is the *sole organ* of the mind; that mind cannot exist without brain; or that there is any necessary relation between the size, form, and weight of the brain, and the degree of mental development. Even in man there is no necessary relation between the size, form, and weight of the brain, and the degree of mental development, while the phenomena of disease in him shows to what extent lesions of cerebral substance occur without affecting the mental life. Physiologists are gradually adopting or forming a more and more comprehensive conception of mind, and are coming to regard it *as a function or attribute, not of any particular organ or part of the body, but of the body as a whole.*" Long ago, the illustrious Milton, discoursing of mind and its seat, properly described the human mind as an attribute of man's body as a whole. In various forms and words, this view has been expressed in recent times by Muller, Lewes, Laycock, Bashman, Bastian, Maudsley, Carpenter, and others. According to these authors, the seat of mind is throughout the body (Muller); mind pervades the body (Laycock and Bashman); mind comprehends the bodily life (Maudsley); psychical life has no one especial centre (Lewes); the whole nervous system is the seat or organ of the mind, the brain being only its chief seat or organ (Bastian). The brain, then, is only *one organ* of mind—the organ, it may be said, only of special mental functions. The old doctrine or assumption of the phrenologists, as represented by Gall and Combe—the doctrine in which they have so greatly prided themselves, and foolishly continue to do so—that, namely, which regards the brain as the *sole organ* of the mind, must unquestionably be given up. We must henceforth regard the true site, seat, or organ of the mind as *the whole body*, and *this is the only sound basis* on which the comparative psychologist can begin his studies. There would be the less difficulty in accepting such a basis, were it only borne in view that the *muscular* as well as the nervous system, that muscular action, has an intimate relation to mental phenomena, to ideas, as well as to feelings. *Muscular action is essential in certain, if not in all mental processes;* e. g., in feeling or emotion, outward muscular expression (*e. g.*, facial) and inward ideas and feelings are inseparably correlated (Maudsley).

I might continue this form of evidence indefinitely, but will mention but one more proof, and this will show how near the eminent anatomist, Sir Charles Bell, came to discovering all the bases of mind. I quote again from Mr. Lewes's "Physiology." Sir Charles Bell, he says, had a *strong*, but *dim*, conviction that it was an error to limit sensation to the action of the special senses. "It appears to me," he says, "that the *frame* of the body, exclusive of the special organs of seeing, hearing, etc., is a complex organ—

I shall not say of sense, *but which ministers, like the external senses, to the mind.*"

Another of this great anatomist's conceptions was that the muscles were a distinct sense. This I have referred to elsewhere. Briefly alluding to this view of Sir Charles Bell, Mr. Lewes remarks:

"Whether it is legitimate or not to elevate this into a distinct sense—a rank denied to the glands and alimentary canal—may be questionable; but there has long been a general unanimity as to the fact that the muscles are the sources of peculiar sensations, such as those of exercise, weariness, cramp, etc. It has also been admitted that the *adjustments* necessary for all movements, for walking, riding, dancing, sitting upright, and so forth, are dependent upon the *sensitiveness* of the muscles. The body is balanced by an incessant shifting of the muscles, one group antagonizing another. But this would be impossible unless each muscle were adjusted and co-ordinated by *sensation*."

Elsewhere Mr. Lewes remarks: "If every distinct part of the organism which is the source of distinct sensation is to be called a Sense, we must necessarily include the *muscles* and *viscera* among the senses; for the sensations derived through the muscles are as *specific* as those derived through the eye or tongue; and the *glandular sensations* are assuredly distinct from those of the muscles. The sensations derived through the viscera and muscles are not less specific, nor less important, than those of eye or ear. We are not at liberty to reject this fact, because it is capable of proof as rigorous as the proof of the existence of Sight or Taste. Mind is the sum total of the whole sensitive organism. No one exclusive organ of mind can be said to exist."

I might bring the strongest proofs of my position, that mind is to be found in the action of the organs of the several viscera and other functions, as well as in the ganglia of the entire organism. I could bring the statements of Haeckel, of Maudsley, of Herbert Spencer—in fact, of all the more advanced thinkers of the world—but the limits of my volume

12

forbid. It is true that no scientist has, so far as I am aware, brought forward the main principles which I here present. It remains for me to elaborate and carry to a finality my theories in my own particular branch of science. At the same time, it is a very great recommendation to my theories that they receive the support (in any degree, however remote) of the best and most advanced thinkers. Although the task of connecting the proofs has fallen to me, it is both a task and a pleasure. It is made the easier for the reason that I have the whole world of living animal and human organisms from which to derive my proofs.

I hope that my readers will make use of the faces and bodies most accessible to them, and those whose physical characteristics they have knowledge of, to investigate and verify the statements and laws here laid down. From these physical peculiarities very much of the mental capacity can be ascertained. The differences in human mind are known by differences in formation of the body, in color, and in inherited quality. Education also modifies natural inclinations, but the forms produced by the arrangement of the internal organs, and the motive apparatus, the muscles and bones, as well as the *colors* resulting from chemical action, are in every case found to be the dominating influences. A few moments each day given to comparison of persons pursuing different vocations will render the observer in a short time quite expert in detecting differences of character, and also in proving the source and origin of so-called *mental* abilities. Let him compare, for example, twelve carpenters with twelve tavern-keepers or merchants, and these with the same number of artists, musicians, and literary persons. Apply the laws herein given in each instance, and I believe that all former errors and ignorance of the character of man will be modified, changed, or entirely dissipated.

As the announcement and demonstration of Galileo's discovery of the laws governing the earth and the heavenly bodies revolutionized (necessarily) preconceived ideas of our planet, its place and rank in the solar system, as well as all popular ideas and beliefs in regard to man, his character and destiny, so the discovery of the facts and laws of physiognomy is bound to revolutionize nearly all existing ideas and beliefs regarding man's character, the origin of mind, religion, art, and the forms and science of government. It must and will change most of the prevalent notions in regard to *woman*, her powers, her rank in Nature, and her ultimate position and destiny on earth.

There is a history attached to the development of every organ and function of the body. There is also a history appertaining to every faculty of the mind. If these two things (faculties and functions) are not correlated, how does it occur that they have kept pace with each other? Why is it that as organs and functions have arisen by differentiation certain "mental" faculties have appeared simultaneously, and always the same functions and the same faculties appear in combination? For example, where we find an excess of muscular power of fine quality, we observe *artistic* and *emotional* abilities; where we find the Vegetative system paramount, we see an utter lack of all literary and artistic conception and ability for those works, but always we find the *functions* of sustentation, reproduction, respiration, etc., strongly exhibited.

The argument and proofs are irresistible to any unprejudiced mind, to every fair-minded thinker and observer. A superficial observer may not see nor accept the proofs, but I think that it may be conceded that a period of thirty years spent in constant research, in this as well as kindred sciences, has not ended in bringing forward vain delusions and chimeras. I hope that nothing which I have written will be taken for granted without evidence. A *reasonable proof* should always be demanded. At the same time, a man who denies what he cannot disprove, or who accepts anything without reasonable evidence, is either guilty of a want of common honesty or possesses a very narrow intellect. Let none of my readers be found in either class. In entering into the consideration of the *rationale* of physical functions and their signs in the face, I give only those signs for organs and functions which long observation warrants me in considering as established beyond a doubt. I am not perfectly sure as to the others, and therefore refrain from indicating them. Time and observation will, I believe, assist in locating them all. The signs of the most important ones are, however definitely located. While proceeding with the consideration of this subject, the reader will please keep the fact in view that the primary use of every function and faculty is for the sustentation, preservation, and perpetuation of the individual; the secondary use is intended (if we may judge from Evolution) for the progress of mankind toward a higher development of the human family.

Let us commence our investigations in the Chemical or underlying division, and analyze the signs in the chin.

CONSCIENTIOUSNESS.—The *width* of the chin, caused by width of its *bony structure*, denotes Conscientiousness; also the strength and action of the kidney system. A narrow, retreating chin shows that the kidneys are narrow and small; a broad, bony chin announces broad kidneys and relative breadth at the "small of the back," as it is termed. By the kidney system I mean, not only the kidneys, but the several ducts and the bladder, as well as all the apparatus included in the performance of its functions; that is to say, all of the *fluid circulation* of the body concerned in the excreting of the *fluid waste*, and the fluid upbuilding of the entire body. Taking into consideration that, inasmuch as seventy-five per cent. of the human organism is composed of water, and the importance of water as a fluid solvent of all the materials taken into the system, as well as its very important office as the carrier of all the materials through the veins and absorbent and secretory tubes, to the several tissues involved in the human organism, it must be apparent that upon the power and activity of the fluid and kidney systems man depends very largely for the purity and integrity of his physical powers; hence, of his moral nature. If the kidney system is not capable of excreting the waste of the fluid circulation, it is thrown back or retained in the body, thus destroying the soundness and integrity of the whole organism; or if it fail, as a common carrier, to convey the particles of lime and other materials needed in sustaining the power of the Bony system; or if the fluid circulation is incompetent to perform its mission in conveying other materials in their right proportion to their several destinations; the organism *will suffer from an unbalanced condition in its physical as well as moral development.* We cannot separate cause from effect; one cannot be moral without the physical powers first possess purity, integrity and equilibrium, in their components and action. Some may object to this showing of the dependence of the moral powers on the physical functions, as antagonistic to Theology. Now, if we could demonstrate morality without the organs and members of the body, this theory would be untenable. If Theology cannot agree with the laws of God as shown through the laws of Nature, so much the worse for Theology.

Morality is related to the use of the members and organs of the body; we cannot be immoral without using them. *We can be neither moral nor immoral in thought alone.* It is by the improper use or diseased conditions of our organs and members that we can become immoral. Morality is not a sentiment at all; it is not a matter of belief or speculation, but a living, actual reality, related to the right use of our physical powers. Almost every individual will admit that certain persons *look more honest or moral* than certain others; also, that some are very sensual looking. The investigation of their conduct often proves that their lives and their looks correspond. Now, what creates this correspondence?—and what causes the difference between moral and immoral persons? Is it the nature of their surroundings alone? No; for, with equal opportunities and temptations, some are able to conduct themselves with more morality than others. Is it not, then, in inherited organisms?—is it not in certain combinations of organs, bones, blood, muscle, and tissue, so placed as to produce certain forms, which, by virtue of these inherited forms, the individual is able to be either moral or immoral? Is it possible for us to know how these moral or immoral qualities are produced, and are we not capable of understanding which forms are most inclined to morality or immorality? I claim that all this can be known; and not only that all these qualities can be detected, but that, by judicious mingling of forms and systems, vices can be bred out, and virtues bred into the human family, just as it is done with the lower animals. What we need to know is, first, the meanings of the several organ-systems and forms of the body; and then a wise and judicious combination of these principles, added to hygienic diet and health conditions, and moral and intellectual surroundings.

FIRMNESS.—Firmness, denoted by length downward and forward of the ramus, or lower jawbone, shows power in the individual to resist disease by the exercise of a firm and persevering determination to recover, as well as the power to persevere in a course calculated to restore health. This faculty being related to the Bony system, indicates that there is organic power—the power which the *conscientious nature* of bone yields—and this is useful in combating diseased conditions. The face of Dr. Tanner (who recently undertook the task of fasting forty days, and succeeded) exhibits this faculty in a remarkable degree. His firmness and perseverance contributed materially to his success, while the superior bony structure which he possesses shows that the kidney system is uncommonly well developed. These two faculties—Conscientiousness and Firmness—will carry one through not only great physical, but also great moral disorders, and enable their possessors to stand upon principle against a world of opposition. Had Dr. Tanner exhibited equal power in other parts of his mind and body, but without Firmness, he could not have accomplished his self-imposed task.

DIGESTION, or ALIMENTIVENESS.— Digestion has its principal sign in the face located on either side of the mouth, and is known by fullness of the lower part of the cheek. This is the most prominent sign, in infancy, of good assimilative and nutritive powers. It is true that the signs of good digestion are to be found all over the person, and the bones will be well covered with adipose tissue where this function is vigorous. There is a seeming want of inductive ratiocination on the part of the majority of persons, who, while they recognize this sign for one physical function in the face—viz., that of good digestive powers—look no farther for the *signs* of the action of the other *visceral* organs, such as the liver, kidneys, heart, lungs, glands, stomach, etc. Now, if Nature has placed the sign for one function in the face, it is logical and natural to infer that others are also represented there. While this function (digestion) is the sustainer of all the mental faculties—that is to say, gives the nutrition essential to their existence and activity—the kidney system keeps all in purity and soundness by its excretory qualities alone. The fact that the fluid waste of the body exceeds the solid waste is undeniable. By actual demonstration, it has been proved that the fluid waste from the kidneys and sweat glands is more, by several pounds weight in twenty-four hours, than is the waste excreted from the bowel system. Writers on Physiology are unable to account for the origin of the sensation of hunger. They do not seem to be able to discover how the individual becomes conscious of the need of the body for more nourishment; that is to say, how the stomach is made to feel hunger.

Dr. C. Cutter, a writer on Physiology, observes: "It has been inferred, by some writers on Physiology, that the glands which supply the gastric fluid, by a species of instinctive intelligence, would only secrete enough fluid to convert into chyme the aliment needed to supply the real wants of the system." What are the reasons for this inference? There is no evidence that the gastric glands possess instinctive intelligence, and can there be a reason adduced why they may not be stimulated to extra functional action, as well as other organs, and why they may not also be influenced by habit. Precisely what Dr. Cutter means by "instinctive intelligence" he does not explain; and, until he gives his explanation, we can find no solution to the question as he propounds it. How does the human system know when it requires nutriment? My theory has been stated before, and I should answer, from its mind! For, as mind inheres in every part of the body, so the branches and filaments of nerves connected with the gastric glands convey to the brain these wants of the individual. The pneumogastric nerve ramifies upon the stomach, and this nerve becomes cognizant of the wants of the organ over which it presides, so to speak; and communicating with the nerves of the other parts of the organism involved in the process of digestion, all combined make demand for more nutrition, and prepare the several organs and glands for its reception. This theory is clearly proved by the fact that, where the brain is functionally or structurally diseased, it is often incapable of taking cognizance of the conditions and appeals of these parts, and insane persons are often compelled by force to partake of food, as they would starve to death if left to their own care, not being notified by the stomach of the needs of the body—the consequence of the diseased condition of the brain. The case of the wounded sailor, noted by Sir Astley Cooper, illustrates this theory, and shows that all the vegetative processes of the body can go on without any assistance from the brain; that is to say, without the brain being *conscious* of the action of the organs of respiration, secretion, excretion, or growth.

In Sir Astley Cooper's "Lectures on Surgery," the following singular case is noted: At Gibraltar, a sailor fell from the yard-arm of a ship, and was taken up unconscious. He remained some months in the hospital there, in a perfectly insensible condition. He was then conveyed to England and placed in a hospital at Deptford, where Sir Astley Cooper, the eminent surgeon, visited him. He was informed by the attending surgeon that the sailor had been insensible for many months. He said: "He lies on his back, with few signs of life; he breathes, indeed, has a pulse, and some motion in his fingers; but in all other respects he is deprived of all powers of mind, volition, or sensation." *If he wanted food, he had the power of moving the lips and tongue, and this action of his mouth was the signal to his attendants for supplying this want.* This last sentence corroborates my theory of the mental power of the nerves of the digestive apparatus. It is here proved that *consciousness* was suspended for many months; yet the organs of digestion had power to manifest *intelligence* in the manner indicated above. This man lay in this condition for thirteen months, when Sir Astley Cooper trephined him; that is to say, raised the depressed portion of the bone from off the brain, upon which it was pressing. Four hours afterward he was able to sit up in bed and converse, and four days after he was restored to all the faculties of his mind and functions of body. He said that he remembered nothing from the moment that he fell; thus proving that the faculty of Memory of Events was entirely suspended. His reason, we see, was dormant; all power over the muscles, with the exception of a slight motion of the fingers, was gone; yet this man lived, breathed, secreted the juices of the stomach, liver, and intestines; excreted from the kidneys and bowels; but was unable to manifest intelligence, except that sort which the digestive apparatus was able to make apparent.

BENEVOLENCE.—Benevolence, shown by the *full, rolling, moist under lip*, indicates a strong and active condition of the glandular system, both of the secretory and absorbent systems. Whenever this feature of the physiognomy is well developed, most of the secreting glands—viz., the lachrymal, salivary, and mammary glands, pancreas, liver, prostate, and testes—will be found to coincide in their vigor and normal action with the size and moisture of the under lip. The absorbent glands also find their illustration in the same feature. "Glands are divided into two classes—the lacteals and the lymphatics. The lacteals are found only in the abdomen.

Their office is to convey the chyle, which they absorb (after the food has been digested in the intestine), to the thoracic duct, whence it is sent into the general circulation to repair the waste and renew the tissues. The lymphatics, on the contrary, are distributed through all portions of the body. Their use is to take up by absorption all waste or useless matters, and convey such matters which have become solvent either to the general circulation, there to be discharged from the system by some of the excretory organs, or are used again in the economy of the human organism."*

I have inserted this slight description of the office of the glandular system, in order that those of my readers who are not well read in physiology and anatomy may understand the philosophy of the action of these glands and the appropriateness of their signs in the face. Now, the glands in the lower lip, being more numerous and more prominent than in any other part of the face, would seem to point to that feature as the facial index of the glandular power of the entire system; added to the fact that the absorbing glands are directly related to the function of digestion, and whenever a prominent sign of any function or faculty is observed in the face all minor signs are always to be found in juxtaposition with it, just as in the body all the organs which assist a similar function or class of functions are placed in positions of sufficient contiguity to facilitate their action. If the entire glandular system is well developed, we must infer that the absorbents will take up sufficient material to supply the necessities of the organism by creating new tissues, and that the excretory glands will perform the task of carrying from the system all effete or waste matter. Hence, a good development of this system shows its power to throw off diseases as well as to resist the approach of those which affect the glands more particularly. Thus you will observe that Benevolence in its developed state assists in protecting the body, as well as gives the power and desire to assist others. We cannot give if we are in an impoverished condition, and

* Harrison's Anatomy.

cannot warm toward others if we are deficient in what creates animal heat. A *thin dry under lip* indicates the reverse of Benevolence, and shows a constricted or impoverished condition of the glandular system, as well as a stingy, close-fisted person.

AMATIVENESS, OR LOVE OF THE SEXES. —Amativeness and reproductive capacity are known by thickness and redness of the centre of the upper lip. When very thick, it also denotes glandular, muscular, and adipose development. This sign is better defined in the physiognomies of ancient races and in European faces than in American people. The function of reproduction is more active with the former; consequently, the sentiment of love, of conjugal affection, of the unity of the family, of the oneness of husband and wife, is stronger and more enduring. These faculties and sentiments being strong and active in the ratio that the reproductive system is well developed, European families are larger families than Americans. This affords greater activity to the reproductive system generally; and as large families have been character-istic of these people for centuries, these sentiments, these faculties as well as functions, have become powerful by use and exercise, and have thus aggregated strength in this di-rection and transmitted it. Hence, its signs are more per-ceptible in the former. This function assists in creative art very materially. The organs of reproduction are mainly muscular, and as this system (the Muscular) includes the power for artistic efforts and creations, the aid which the reproductive system offers to creative efforts in art, etc., is only added proof of my theory of Amativeness. This the-ory can be proved by investigation of the countenances of all of the most talented persons known to history. Examine portraits, for example, of Dickens, Moliere, Byron, Sterne, Lady Morgan, Hood, Albert Durer, Cervantes, Shakspeare, Burns, Shelley, Joanna Baillie, Sarah Siddons, Lavoisier, d'Alembert, Rousseau, Buffon, Des Cartes, Delambre, Ark-wright, Jeremy Taylor, Murillo, Corneille, and Joshua Reynolds. In fact, the examination of the faces of all who have excelled in *creative art* will disclose the sign of this function and faculty well defined.

Its use primarily is for the propagation, creation, and perpetuation of the race. Its *moral significance* is of incalculable importance, for upon its *normal action* and natural and religious use the purity and welfare of the human family are dependent; It has no functional activity until the age of puberty, at which time important moral as well as physical changes occur. These changes are equivalent to the introduction of an entirely new faculty and function. Its full moral and physiological importance should be taught to youth, as ignorance of the true nature of its powers may lead to disastrous results, which may descend to the innocent for generations and lead to the utter demoralization of entire communities.

LOVE OF CHILDREN AND ANIMALS.—Love of children, pets, and animals is known by the drooping of each side of the upper lip on either side of Amativeness, of which it is the natural and necessary companion. It forms a little "scallop" shape, which also assists in giving beauty to the mouth. Indeed, all well developed mouths present this appearance more or less. Every function that is of use to the individual, and in a normal condition, sets a sign of beauty in the face; and those who learn to understand these signs and their signification will enjoy beauties which are denied to those ignorant of them.

In some the outer sides of the lip project downward almost overlapping the lower lip, just as we see it in dogs and cows, and other animals whose love of offspring is intense. This sign is located in the same place in all the higher animals. As I have stated elsewhere, when Nature gives the love or capacity for any pursuit she also gives some kind of power for its expression. Hence, when we observe this sign largely defined, we must infer that the ability to nourish or care for the young accompanies it. In some it betokens the physical development essential to the nourishment of offspring; that is to say, good digestion and a suitable endowment of the glandular system. In others, in the brain system predominant, it is accompanied by a mirthful-constructive ability, which manifests itself in the invention of stories, games, and amusements for the diversion of the young. Miss Louisa Alcott, the celebrated writer for children, exhibits this formation; all the signs of this kind of talent are prominent in her physiognomy.

This faculty is manifested in others by love of teaching and training young children and animals. No one can succeed in training dogs or horses who has not this faculty. I have recently studied the portraits of our National Fish Commissioners and their assistants. Most of them show this sign well defined. In order to be in harmony with their work (the propagation of the finny tribe), the two faculties of Amativeness and Love of Young should be present, so as to facilitate the work of reproduction and care of animals. All of the faculties and functions in the Vegetative or Chemical division of the face are related in some degree to the glandular system. Now, as love of offspring is generally stronger in woman than in man, she is by Nature especially fitted to nourish the young, and the *sentiment* of Love of Young is created and sustained by the glandular system—by the mammary glands in particular. In man, these glands are rudimental; hence, his love for and desire to nourish and take care of the young are not so strong as in woman, although several well authenticated cases are found in medical works of men who were able to nourish babes at their breasts. There are a few ducts and a small gland in the mammæ of men, it is true, and it is quite likely, under some abnormal conditions of the generative function in man, that the mammary glands have become enlarged, as is well known in cases where the testes have become atrophied.

This function and faculty, it will be observed, has its moral and intellectual use, as well as its physiological power. It is, therefore, highly important as being one of the greatest protectors of infant life and health and the conservator of posterity. The functions and faculties in the Chemical division of the body are the most easily recognized by the ordinary observer; more profound thought and reason are necessary to carry this law of correspondence of *functions with mental and moral faculties* to its ultimate conclusions.

MIRTHFULNESS.—The most prominent sign of this faculty is found at the outer corners of the mouth. It is shown by a depression caused (when smiling) by the action of the two muscles named major and minor zygomaticus, which draw the mouth outward and upward. The more these muscles are exercised the more defined the impress of such activity is apparent, and hence it is that we often find dimples at this place. In those less playful and mirthful, small vertical wrinkles are seen. This sign adjoins the local sign for Love of Young, and is connected naturally and necessarily with it. In some it causes the corners of the mouth to turn upward. Laurence Sterne, the celebrated humorous writer, has this peculiarity in a marked manner. It is adapted to the care and amusement of the young as well as to the recreation of adult life. It is in one sense constructive, like Amativeness, as it assists in contriving and planning amusements for old and young; it shows in witty and funny speeches, and attracts all by mirthful and lovable manners; it is also an aid to digestion, and adjoins its most prominent sign. All display of anger or sadness while eating impedes digestion, while mirth assists its action. Its source of supply is undoubtedly glandular, although the muscles, too, assist in its expression. The zygomaticus minor muscle is sometimes scarcely perceptible or entirely wanting.

The location of Mirthfulness near the mouth, and its intimate relation to Love of Young, point to its origin as glandular, depending undoubtedly on the quantity and quality of nutrition assimilated and animal warmth supplied to the system by the action of the lacteal glands. Shriveled, thin persons, or dyspeptics, are not as mirthful as those whose digestion is unimpaired; and as dyspeptics regain health and normal conditions their love of fun and mirthfulness return to their natural state. The location of this function and faculty, and the effect of its normal and abnormal action, evidences its origin. Like all the faculties found in the Vegetative system, it must be considered as having its support from sources similar to those of other functions and faculties in this system. The association of all these functions is for mutual support and assistance; hence, their origin is easily determined. To "laugh and grow fat" is a truism. Anger and sadness suppress the normal supply of secretions, while mirth and contentment excite them to action.

FRIENDSHIP.—Friendship is related to and sustained by the intestinal system, and is comprised in the chemical or vegetative part of the process of digestion. Its principal local sign is fullness of the upper portion of the cheek, and adjoins the chief sign for Digestion, or Alimentiveness. Fullness of the salivary glands just in front of the ear opening is another sign of the assimilative capacity. The first stages of digestion—those performed by the stomach—are produced by muscular action chiefly, with slight assistance from the chemical action of the salivary and gastric juices. The most important part of digestion is carried by the alimentary canal, commencing with the duodenum. The food, in its passage through the intestines, is acted upon by the secretions of the liver and pancreas; and in this part of digestion the process is mainly chemical; and it is here that the juices needed for animal heat and warmth, for the nutrition of the body generally, are found. It is here that *color* is evolved by chemical action, and sent through the glands and veins to its several destinations in the tissues, by the power of the same action, without the slighest assistance from the Muscular system; and when we observe fullness of the upper part of the cheek, of a bright red color, we know that Friendship is active, because the power, the warmth essential to its action is present in the body in the right proportion to enable the individual to perform the offices essential to the active duties which Friendship exacts. A thin, flat, pale or bluish upper cheek shows the reverse of this faculty, and will always be accompanied by a small or defective bowel system.

Friendship, like Love, is at one and the same time a benevolent and a selfish trait. Its character is dual, as is its functional action. It both secretes and absorbs. Primarily, it seeks to *please itself* in social enjoyments, in the society of friends, and in eating and drinking with them. It is not, like the Irishman's "reciprocity," all on one side. It seeks, also, the enjoyment of those it loves; and, where there is a good admixture of the architectural or mathematical powers, it assists, by planning and personal service, in every way the interests of the objects of its affection. A good development of the bowel system gives to the organism the juices and nourishment needed to carry forward the work of friendship, and also to give the animal warmth essential to the creation and perpetuation of this faculty, either as a sentiment or social enjoyment. Its physical basis is, as I have shown, in the chemical division; and, in its primitive aspect, it leads to desire for association and companionship. In the early stages of man's development, it assisted in forming tribes and clans, and the faces of all clannish races exhibit this faculty largely; as, for example, the Highland Scotch, the Swiss, Hollanders, and others. As the organism rose higher by the development and perfection of other faculties, it exhibited itself more as a *sentiment*, and showed its action by pleasant speech, in thought, care, and active works. In combination with the Chemical division, it will show by entertaining friends with feasts, by cooking for them, and by presents of nice foods, drinks, and by attention to their bodily wants. With the Architectural added, it shows in entertainments, also, but adds both *sentiment* and good deeds. With the highest, or Mathematical division large, where the brain and nerves impart sensitiveness, it will be exhibited more in emotion, feeling, thought, and sentiment;

in poetry dedicated to beloved objects; by presents of flowers, books, and pictures, and delicate attentions.

The Germans, as a class, are the most sociable and friendly of all the civilized races. They are also the best *feeders*, with most uncommon assimilative powers. Hence, it will be seen that friendship is a conservator of life, and assists in the progressive development of the human family, both morally and physiologically. Some of the glands involved in digestion are both secretory and excretory. This dual action gives rise to dual action in the manifestations of friendship; they are both selfish and unselfish.

MODESTY.—The most prominent sign of Modesty is shown by a vertical depression running down the center of the upper lip. It is an unfailing sign of a love of purity, cleanliness, and generally of chastity; all of which are conducive to health and long life. Persons exhibiting this sign love refined language, dislike all coarse or smutty jokes or allusions; love neatness of attire, and desire to change their clothing often; dislike bad odors emanating from the teeth or skin; bathe frequently; and in all ways testify to chaste and modest tastes. Its location near Amativeness suggests the beauty and utility of its placing.

CAUTIOUSNESS.—One of the principal facial signs of caution is shown by extreme length of nose. Its principal use is to protect the body by its sense of scent, which prevents all hurtful and noxious materials from entering the stomach, and keeps poisonous gases and odors from the lungs. The sense of scent acts as a sentinel; hence its position, directly above the mouth. This sign is conceded by Simms and Fowler. In the animal world this faculty is more used than in the human race; we depend more upon our eyes and acquired experience. The eyes and observation not being so well suited to this purpose in animals as they are in man, hence it is that all animals smell their food constantly during a meal. The herbivorous animals, while in a natural state,

13

seldom touch any grass or herb which is poisonous or detrimental to them—so unerring is their scent; yet, after becoming domesticated, they lose this sense partially. This sense is at least as high as man's power for observation; yet people usually speak of it as "animal instinct," giving the idea that this faculty is something inferior to human observation, while in reality it is far superior to it; for no human being can tell by scent alone, without experience, whether certain plants are hurtful or useful. In many directions animals possess superior powers. Had they a suitable physiological development which would enable them to speak, they would soon convict us of more cruelties, meannesses, and contemptible behavior than even wild beasts are guilty of.

VENERATION.—This function and faculty is related to the stomach, which is our creator physically. Height and width of the "bridge" of the nose is its principal local sign in the face. Unlike the intestinal system, its functions are mainly mechanical; it is, therefore, located in the Architectural or Mechanical division. The stomach is the receiving laboratory, where the solid materials are first mixed by mechanical action mainly. This operation is called peristaltic motion, and is produced by the contraction of the muscles of the stomach, and the expansion and contraction of the lungs and diaphragm; the saliva and gastric juice performing only a small part of the chemistry of digestion. The materials taken into the stomach, after being thus acted upon, are distributed for chemical action, which must be performed before the act of creating and replacing new tissues, bones, etc., is completed. Although we are dependent upon the fluid circulation to convey to their destination, in the liquid form, all the materials for the maintenance of the body, at the same time suitable solid materials must be furnished to the stomach, to be by its mechanism converted into chyme, a kind of pulp. Thence its further progress is continued to the duodenum, where it attains a fluid state, called chyle; this is received into the general circulation, and assists not only in nourishing the body, but it also gives the materials essential to the creation of other human organisms. The *height* and *width* of the nose at the bridge (or where the sign of Veneration is found) is the indication of the power of the stomach to do its work. The low or "scooped" nose is always accompanied by a predisposition to weakness of this organ. Weakness of the stomach does not necessarily involve weakness of the bowel system, as one depends upon the muscular power, and the other part of the process of digestion—the chief part—upon chemical action. Over thirty feet of intestinal surface (according to physiologists) are traveled over before the process of digestion is complete.

In the animal kingdom, we find that in those who have very flat noses, such as monkeys and apes, and all flat-nosed creatures, dyspepsia is quite prevalent—more so than with camels, horses, dogs, elephants, and those with a higher nose. Dyspepsia leads to consumption, which cuts off both men and animals who exhibit this formation of the nose. Persons and animals with long, slim necks are also predisposed to consumption and dyspepsia; and, accordingly, we find giraffes especially subject to dyspeptic ailments, even in their natural state.

Hope.—The degree of this very important faculty found in an individual is dependent upon the normal action of a strong and healthy liver. If the liver is of good quality—that is to say, free from all inherited weakness and always acting normally—a high quality of Hope will accompany its action. Hope is a great sustainer of life; it buoys one up under great difficulties; it gives power for overcoming obstacles by a hopeful, cheerful cast of mind—if I may be allowed to use this term in speaking of a physical function, for we derive our "mental powers" from these functions direct. In sickness no faculty except Firmness so sustains the spirits and strength of the invalid. In this way it promotes health and longevity. Whenever I see an individual with cheerless, despondent, hopeless views of life and the future, I look for a liver diseased either by abuse or by inheritance from some "blue," grim, joyless, jaundiced, bilious ancestor, and I find this invariably the case. How little people think, as they stuff and gorge and make themselves bilious and jaundiced, of the gloom and wretchedness they are storing up for future generations, cursing the unborn, and sending down to posterity the blighting effects of their uncontrolled appetites. Surely, it is here religion should commence where it is most needed; and Nature has placed Conscientiousness in the Vegetative division in the physical basis of human character, in order that it should protect the body in purity and soundness, and that morality should prevail.

When I see persons whose views of life are gloomy, and who live without hope, I cannot refrain from paraphrasing the Scriptures thus: "The fathers have chewed gall, and the children's teeth are set on edge." I suspect there must have been many keen, observing, thoughtful men in "Bible times," who were wiser and more scientific than they dared to acknowledge—some who understood, as Moses did, the physical construction of the body, as well as man's requirements toward a religious life. When I read such expressions as the "gall of bitterness," "bowels of mercy," etc., I cannot but think that some of the men of those times must have known that friendship derived its merciful attributes from the bowel system, and hopelessness and bitterness of spirits came from an overflow of the gall-bladder; else why such expressions? In those days a man who "knew too much" was called a "sorcerer;" in these days, if he dare mention the "bottom facts" in regard to the operations of God's laws as exhibited in Nature's works, he is generally assailed with the opprobrious epithet of "infidel" or "materialist." Yet how any one can demonstrate the existence of God and his laws without material substance, I am loss to understand. As burning and stoning do not follow such expressed opinions as formerly, they are allowed to exist. When I see attacks made upon those giants of science, Darwin, Huxley, Spencer, Haeckel, and others, because like Galileo they have discovered greater truths than their petty ignorant opposers can comprehend, I rejoice that they live in the nineteenth century under the reign of the civilization of the printing press; for were we still under the dominance of the ecclesiastical powers like those of the middle ages, the faggot and torture would be their portion. The probabilities are that these great scientists will live to be instrumental in disseminating such knowledge of the laws of God as will assist materially in the advance of a high civilization in spite of pope or priest, bigot or ignoramus.

ANALYSIS.—As I have previously shown that Hope derives its power from the glandular system—viz., the liver—so also we will find that analytical power is in strong sympathy with the same organ. Its sign and location adjoins that of Hope, and both are at the end of the nose, directly under the cautionary action of the nostrils. These two faculties and functions (Hope and Analysis) occupy a position about midway between the Vegetative, or chemical, and the Muscular, or mechanical, divisions of the face and body, and are both assisted by the action of the liver. This organ has the power of excreting and secreting, and assists by its clearness of action in the so-called mental operations necessary in mechanical, artistic, and literary work.

As we ascend in the scale of development from a lower to a higher grade—from the Vegetative to the Thoracic, from the Thoracic to the Muscular functions—we find these different growths overlapping each other, as it were, and this peculiarity is noticeable in every department of organic life. This method is not only apparent in the successive growths of the same organism, but it is also very marked in the evolution of species, where we often see, as in the amphibia, functions which are useful both for aquatic and terrestrial existence. So, in the human family, as I read in the face and body, we see the remains of former existences, the remains of our animal ancestors. Not only are these inheritances characterized by phenomena which the popular voice terms "animal passions," such as hate, revenge, destruction, and jealousy, but we see in the uselessness or purposeless of numerous organs, and parts of useless members, which are scattered in different portions of the human organism, the greatest proofs of evolution, and indeed the one of all others which would establish the truth of that doctrine on a firm and unassailable foundation. I shall give brief mention of these rudimentary remains of our ancestors, referring the reader to "The Evolution of Man," by Haeckel, for further light on the subject.

The generality of people accept, without question or analysis, the human organism as they find it, never glancing back to trace laws and appearances to their origin; but as soon as inquiry and investigation commence, light flows in upon them. I will refer to only a few of the remnants of former animal existences that are now found incorporated in the human system, without having any use or purpose in man's economy, but which, on the contrary, often induce disease and suffering. I will first call attention to the fine short hairs found all over the body. What is their use to man? None, that physiologists have been able to discover. They are simply vestiges of the thick hairy covering found on our most immediate animal ancestors. The little circle of muscles surrounding the ear-shell is another relic of an existence which found flapping and raising and lowering the ears a necessity. These muscles are not of the slightest use to man, as his ears are immovable. At the inner corner of the eye we find a little fold of skin, the remains of the nictitating membrane, the third eyelid, which is useful to some birds and fishes, such as owls, sharks, and others, but serving no purpose in the human family. At the termination of the vertebræ, or backbone, we have five little bones, with joints and shrunken muscles, that are of no use to man. They are subject to disease, and I have recently heard of the successful amputation of the coccyx, as this rudimentary "tail" is called. Another useless, and worse than useless, relic is the thyroid gland. It is situated in front of the larynx, and is the remnant of the crop so useful to our animal ancestors. It is often the seat of disease. The swelling so common in the mountains of Switzerland called "goitre" is an affection of this gland, which has no use in the organism of man. There are various other parts, like the vermiform process in the intestines, which are only a detriment to us; also, some atrophied muscles in the thighs, which are useless in our present state of existence. They were very useful formerly in climbing trees, a process our animal ancestors found essential to their welfare. There are many remains of former conditions of the reproductive system when organisms were bi-sexual, and thus it is that in man are found portions of rudimentary female organs that are functionally active only in woman. These have no use in the body save to enlighten us on the subject of our pedigree and descent, and also to teach us the methods of Nature in evolution; yet all serve to illustrate the power of God, who from so small a beginning as a simple germ-cell can create by successive steps the complex being we call Man.

No portion of the human system acts independently, but all of the five superior organ-systems are so correlated that neither can act without being affected by or affecting the others. These several powers are diffused, as it were, through the entire body, although there is a sufficiency of connection and similarity of action in each to enable us to trace its cause and operation through the entire organism. At the same time, each system extends its influence (as we rise in development) forward into the next growth, and there are faculties and functions which seem to belong to and affect the operations of functions and faculties in the next system above; as, for example, mouth-breathing represents the condition of the amphibia before true lung-breathing was established. Yet breathing through the nose is the more perfected method of respiration. The faculty of speech is not essential to the lower systems of the body—the Vegetative and Thoracic, for instance; for we know that all the really necessary operations of life can be carried forward without speech. But as the

functions and faculties become more complex, and art and
reason assert their presence, then speech becomes essential
to the perfection of character. Accordingly, we find all the
requirements of this faculty in the Muscular system, and it
is in this system that a marked advance of functional activity
is made, as it evolves in the regular order of progressive
growth in the lower organisms.

The parallel holds good in man; for where we find the
Muscular system pre-eminent in him, we find ability for
artistic conception and execution of various kinds. Muscular development of fine quality gives the desire and ability
for motion, for architectural, artistic, and other works, which
the lower systems of faculties and functions are not capable
of producing. The sign for Hope, or the liver, in the face
is situated just above the Vegetative division of the physiognomy; yet it seems to assist the action of both it and the
and the other divisions above, particularly the lungs and
heart. We know that this is the fact physiologically; and if
physiologically, then of course the "mental" character is
affected by such interaction. The kind of analytical power
which the action of the liver gives rise to is better adapted
to the analysis of art, literature, mechanism, and science
than the sort which is essential to abstract reasoning; hence,
we observe with inventive, fertile, imaginative, and artistic
persons this sign is very pronounced. The septum of the
nose will be seen projecting downward, with an unusual
clearness and brightness of the eye and skin, thus evidencing
that the biliary system is doing its perfect work.

An excess of Cautiousness conduces to a constricted state
of the liver, and prevents its healthy action. Where this is
the case, artistic analysis and hope are never strongly developed in the individual. Where there is a good degree of
muscular development, combined with Hope and Analysis,
we find the most active people. Persons who are round in
their forms (made so by excess of muscle and a suitable proportion of adipose tissue) are constantly in motion. They
become acrobats, athletes, artists, actors, musicians, and in

business they are speculators; for in the Muscular formation we find the *imaginative* class as well as the artistic, and this induces *faith*, *belief*, and superstition, love of the miraculous and wonderful.   Men with this endowment are inclined to enter into extraordinary methods in business, and are disposed to believe in the existence of fabulous resources, such as mines and miraculous and impossible inventions and discoveries.   Thus believing, they extend their faith, and draw others into their schemes, which are generally outside of the legitimate channels of commerce.   Such persons are usually believers in clairvoyance, in the power of some to foresee the future, and often consult them on business affairs.   They are strong believers in superstitions and improbable religious dogmas rather than reasonable moral ones.   This class of persons are always willing and anxious to be *saved*, but prefer to rely on some power other than their own to save them, and do not care to make great and continuous efforts to lead moral lives, as this requires great self-control and self-denial. The *rationale* of this is simple and easily explained.   Muscle is *yielding* in its nature, like fat, and is unable to be firm and steadfast in *one* position, like bone, which can always support itself, can overcome fat and muscle, and resist pressure from the outside.   Hence, bone is more capable of assisting moral efforts, because possessing more power of resistance. When we desire to know the nature of anything, we must look to its components and method of action.   Scientific analysis can carry this process to its ultimate, and trace the constituents of muscle, tissue, and bone to their origin, and thus give the clue to the mainsprings of human action, of morals, art, and religious systems.

THE LUNGS.

In diagnosing the power of the lungs, the first thing to be taken into consideration is the size of the nostrils.   Generally speaking, the larger the nostrils the greater the size of the lungs.   If the nostrils are large and round, the lungs will present the same peculiarities. If the nostrils are narrow and long, the lungs will correspond in formation. The strength of the lungs will depend upon their inherited quality, *regardless of size*, although the large round lungs are usually the stronger. The strength and power of the lungs may be known by a healthful color of the skin, as well as by a healthful brightness and clearness of the eyes.

In deciding upon possibilities of lung power, the condition of the digestive system must be taken into account. Where assimilation is easily performed, the lungs will be well supplied with good blood; but if the nostrils are narrow, the skin pale or blue, and the cheeks thin or hollow, great care must be taken to provide the stomach with the most nourishing food, else that dread scourge, consumption, will make its appearance. Persons with weak digestion set little value on food, and often neglect themselves in this respect, and in this way the lungs become impoverished and soon decay. Such persons should make a business of eating, and cultivate the appetite by eating all that the taste calls for. Appetite can be cultivated, just as any other defective function or faculty. Poor feeders do not have so strong a hold upon life, nor are they as capable of friendship, as those who nourish the body well. Consumption can be cured in its first stages by pure air and a dietary suited to the individual. *Medicine cannot cure it.* It may sometimes mitigate the severity of the cough, but medicine cannot supply good rich blood in the right proportions. Nothing but good food made into blood, and this blood oxygenated by the purest atmosphere, can replace the diseased and worn out tissues. Medicine never created either blood or tissue; food and air alone perform this miracle.

THE HEART.

When one wishes to discover the size and power of the heart, he can readily find its index in the face. This may seem a novel way to learn the condition of an organ so deeply hidden in the interior of the body. But the reader has already become convinced, I opine, that one of the uses of the face is to disclose bodily conditions; and when we wish to know the size and power of the heart, we must look to the size and quality of its nearest companion and faithful ally for the information. The lungs work in harmony with the heart; they are correlated in their action and mutually condition each other. It would be as great an anomaly to find large lungs and a small weak heart as it would be to find large Love of Children and small Mirthfulness. Nature does no half-way work; Nature is no bungler, for wherever we find a strong digestive system, supplying a bountiful quantity of blood to the heart, we invariably find large lungs to receive it. And so we may set down as a law that the larger the nostril the greater the size of the heart. The cause of this is based on the great quantity of air which large lungs will be compelled to use in oxygenating. This air supply gives greater activity, not only to the heart, but also to the brain. The larger the lungs the more active the brain will become; hence, the entire circulation will be more rapid or stronger. This activity of the circulation will facilitate digestion, for as the brain calls upon the stomach for more food to sustain its activity, the stomach will respond by demanding more nourishment. This demand being met, the body will be kept well nourished and at its normal degree of heat. The hands and feet will always be warm, and the circulation will conduce to make the entire surface of the body comfortable. All this can be seen at a glance, judging simply from the size of the nose. The surface circulation being perfect, and supplied with well oxygenated blood, the complexion will testify to the perfectness and soundness of the lungs and heart by a heightened color of the skin and brightness of the eyes. Weakness of both heart and lungs induces a sluggish, feeble circulation and a pallid or bluish cast of complexion.

COLOR OF THE SKIN, EYES, AND HAIR.

The complexion and color of the skin, eyes, and hair have a moral as well as intellectual and physical signification.

Where the organism is deficient in the coloring pigment (as I have explained elsewhere) a weakness of the glandular system is usually indicated. This deficiency is shown by milk-white or very light eyes, weak hair, and skin of a pallid hue. This appearance is often accompanied by imperfect vision, deafness, tubercles, a scrofulous diathesis, chlorosis, white swellings, and many other diseased conditions of the glands in various parts of the body.

Now, if the sight or hearing is imperfect, the individual cannot gain correct knowledge of material objects, nor of speech and ideas. Persons with defective senses fail to apprehend the perfect and entire import of what occurs about them; hence, they are liable to take in erroneous or partial understandings of things as they appear. They are also, by reason of such defective senses, less able to perceive and avoid dangers, and, by reason of their weakness, less able to resist the attacks of disease and more liable to be affected by immoral temptations.

The glandular system being both absorbent and secretory in its nature, assisting by absorption the function of digestion, would fail in case of defective action to absorb and convey the materials essential to supply the coloring pigment which the foods extract from the minerals contained in the earth upon which they are grown. The glands would also fail in the chemical action necessary to furnish new tissues and animal heat to the organism. Unless all these operations are perfect, Friendship, for example, cannot exist in its highest state. If the secreting glands, the lymphatics, are too weak to properly perform their office, and fail to absorb the impurities of the system, the body becomes charged with waste matter, and a condition of moral impurity will be the result. Can it be doubted, by any observant or logical person, that a sound and pure body is more capable of morality and integrity than one which is weak, diseased, and impure? There must be equilibrium in the several functions of body. and faculties of mind in order to produce harmonious conditions of the moral and mental faculties. The more I investigate the human organism the more I am convinced that the moral nature is dependent for its purity and strength upon physical conditions, and not upon theories, beliefs, or dogmas, although cultivation of the moral sense is necessary for the progress and preservation of the race.

There are many other ways in which the moral and mental faculties are made to suffer by absence of coloring matter. Its deficiency causes people to be suspicious. Lacking the warmth essential for great friendship, they are ever ready to suspect their friends. Ask any very light-eyed person if this is not characteristic of them. A candid answer will prove this statement. On the other hand, too much coloring pigment induces another class of diseases, and evidences other moral and mental peculiarities and defects. Persons with very dark skin, hair, and eyes are liable to disorders of the biliary system, to fevers and inflammations. As with great depth of color intense heat is always found, so we must infer that the passions and emotions of very dark races, such as love, jealousy, hatred, revenge, and the like, are more violent, intense, and heated than those of the white races. This fact is well illustrated in the negro and Indian, as well as in the Spanish, Portuguese, and Italian races, and the inhabitants of the tropics generally. Sufficient color is a necessity and a preservative of life and health. It also gives tone and strength to the moral and mental faculties. Too little coloring pigment, as I have shown, renders the individual weak, morally, mentally, and physically, and induces shortness of life. The knowledge of these facts should be an incentive to the study of hygiene, and the application of its laws to the human organism. Diet suited to each individual, proper exercise and clothing, with sunlight, pure air and water, should be considered as first in the scale of human necessities, and the effort to procure them the first and highest of *religious duties*.

THE MUSCULAR SYSTEM.

The eye is, perhaps more than any other feature, the exponent of the muscular power of the entire body. It contains within its small orbit more muscles, more active and finer ones, and those which express more emotions, sentiments and mental energies, than those of any other part of the muscular system. Hence, the eye is properly the most prominent facial sign of the Muscular development. The larger and fuller the eye and its orbit, the greater is the predominance of the Muscular system in the body. This law obtains in the animal kingdom as well as in the human family. Those animals that are best developed in their muscular systems, and which depend mainly upon their activity for their livelihood, have larger eyes than those which have other systems of functions dominant. The high-flying birds, the vultures, falcons, and eagles, the animals that live in mountainous countries, such as the deer, springbok, ibex, gazelle, etc., have larger eyes in proportion to their size than the domestic animals, the horse, dog, sheep, and camel, in which the Osseous system prevails.

The class of individuals who are engaged in artistic pursuits, such as sculpture, painting, and acting, and who are writers of poetic and imaginative works, are invariably large-eyed as compared to mechanics, scientists, and moralists, and an examination of their organisms will show that the Muscular system is in excess of the Bony. Were this not the case they could not express emotion, for bone is incapable of such expression; nor could they excel in works of art, for in this department of industry the muscles must be able to *take command of the bones*. The sculptor, painter, and actor must be made of more pliable material than bone; so also must the athlete and the musician. They must also be endowed with more active material than fat, for too much fat creates inertia, and is not as intelligent as muscle. Muscle alone is capable of expressing emotion or feeling. Bones cannot do this, neither can fat nor brain; fat is for warmth, for social and sensual enjoyments; muscle is for action and emotion, and brain for abstract reasoning. If brain is for the expression of art, sentiment, sociality, and feeling, as many believe, for what purpose are the muscles and the internal viscera? The phrenological idea that the eye is the "organ" of language, and that the brain back of it pushes it forward, making it appear large where language is well indicated in the individual, is like many other phrenological ideas, only a half truth, like the notion that the sign for the sense of weight is caused by a bulging out of the brain where this sign is located in the forehead. The signs for Weight and Language are caused, not by any particular fullness of brain-matter anywhere, but are both due to the same cause; viz., an excessive development of the Muscular system. We cannot use brain for language; one portion of the brain sits in judgment upon our speech, but neither brain nor nerve-matter can produce one word. It is true, the nerves assist all the operations of the body and mind, as intelligence or mind inheres in every portion and organ of the body; but brain and nerves are not the principal actors in either speech, sentiment, emotion, or art. The action and power of the brain have been entirely overrated, and Phrenology has assisted this erroneous popular idea by setting forth that the brain is *the organ* of the mind. This has been so generally accepted that now even disbelievers in that science have an idea that a large head, and particularly a great bulging forehead, are evidences of a mighty intellect. Nothing can be farther from the truth. Great foreheads are in many cases, if not in most, evidences of great slowness and stupidity, while men with foreheads rather low, and sometimes slightly retreating, are found among the foremost men of the world. Such men are the leaders in thought and action. The physiognomies of Fox, Priestley, Marlborough, La Place, Gambetta, Fremont, Lord Chatham, and many others, who have led in thought and activity in practical life, show foreheads not high and slightly retreating. The sooner we can discard the idea that the brain is all-powerful, and adopt the fact that all parts of the body contribute to the general intelligence and to the phenomena called "mental operations," the sooner will we arrive at a just and truthful understanding of the science of mind and of human character.

The eye has for the carrying forward of its work the assistance of several transparent fluids, as well as a coloring pigment. Where this pigment is lacking in depth of color, as in the case of albinoes, whether it be in human or animal albinoes, the vision is defective. Where the eyes are very light in color, the sense or understanding of the shades, hues, and harmonies of color is wanting to a degree. The retina cannot reflect the image correctly if the pigment is not dense enough, or if it be too light colored, any more than can the camera obscura exclude the superfluous rays, and fasten the picture upon the plate where its color is deficient.

The eye, more than any other feature, unfolds the character of the individual. The form, size, color, quality, and brightness, the manner of placing it in its orbit, and its relation to the surrounding parts, together with that indefinable something which we call expression, and which is incapable of scientific description, must be "sensed" by the intuitive reader of physiognomy if he would comprehend the individual under inspection. We are apt to give the sense of sight far more credit than it deserves. Much of the knowledge which we think we have gained from sight alone is due to the co-ordination of the sense of touch with sight. Let the reader experiment by shutting the eyes, and feeling all the articles in common use, especially by touching clothing hanging in a dark closet. He will find the sense of touch so exquisite that he will cease to marvel at the achievements of the blind.

In analyzing the action of the Muscular system, regard must be had to the differences existing between the flat thin muscles and the round full muscles. These differences originate two classes of character. The thin flat muscles are found in those organisms devoid of juices; consequently, the emotions are not as spontaneous, and the execution of art is not as easy, as when the muscles are full and round. This difference may be known by the eye. The eyes of those with thin flat muscles are large, but not full and protruding, while the round muscles show full prominent eyes.

THE EAR.

The ear, as well as the eye, is an exponent of the Muscular system. In the lower animal organisms no appearance of anything like an ear was found until the Muscular system was evolved; and as the same methods of evolution were employed by the creative power in developing the functions and faculties of the human race as were used in evolving the lower animal races, we accordingly find the ear and all the appurtenances involved in its office in the Muscular division of the body. Without going into a minute anatomical description of the ear, which the reader can do for himself by consulting any author on the subject, it is sufficient to note that every portion of the sense of hearing, like the faculty of speech, is performed mainly by the muscular, cartilaginous, or membraneous structures. It is true that here the nerves assist, as well as certain fluids; but in deciding on the degree of muscle in an individual, we can use the ear as well as the eye for this purpose. If the ear be round and thick, the Muscular system will be well defined, or if the ear be large and thick; but where the ear is thin, the Osseous, Thoracic, or Brain system will be indicated. The ear shows the power of the tympanum to receive sound. The manner in which it is conveyed to the auditory nerve and its terminations is most interesting. I take pleasure in quoting from Mr. Tyndall's excellent work on "Sound" the following description of its production within the ear. He observes :

"The sound of an explosion is propagated as a pulse, or wave, through the air. This wave, impinging on the tympanic membrane, causes it to shiver; its tremors are transmitted to the auditory nerve, and along the auditory nerve to the brain, where it announces itself as sound. We have the

14

strongest reason for believing that what the nerves convey to the brain is in all cases motion. It is the motion excited by sugar in the nerves of taste, which, transmitted to the brain, produces the sensation of sweetness. The motion here meant is not of the nerves as a whole; it is the vibration or tremors of the molecules, or smallest particles. Different nerves are appropriated to the different kinds of molecular motion. The nerves of taste, for example, are not competent to transmit the tremors of light, nor is the optic nerve competent to transmit sonorous vibrations."

A very eminent oculist informed me that perfect eyes are so rare as hardly ever to be met; that is to say, perfect in their mechanism. To hear the different versions of several persons who have listened to the recital of the same narrative, we would infer that the ear partook of the same imperfection. The several senses are all indicative of the general tone and equilibrium of the mind. If an organism be well balanced, I believe all the senses will be more nearly perfect than where an unbalanced condition exists. This should teach us the importance of striving to produce and transmit an equilibrated condition, in order that our sight, hearing, smell, taste, touch, and memory and reason, may be kept in their highest vigor and power, and thus our knowledge of the outer world will be based on the truth—upon reality. The faculties of seeing and hearing are primarily for the protection and preservation of the body. Full protection can ensue only by careful and hygienic treatment. It is wrong to prostitute the eyes by unlawful use, as it tends to weaken our capacities, and destroys our power for receiving correct impressions of what transpires.

WEIGHT.

The physical use of the sense of Weight is primarily to assist in balancing when walking; also, to prevent the individual from falling when exposed in dangerous positions on mountains, high buildings, ladders, steeples, etc., and to balance him in running, leaping, and in all ways where the muscular sense is needed to protect the body. The sense of weight is related to the Muscular system. Where it is well developed the muscles called *corrugator supercilii* will be found much enlarged. (See sign for Weight.) Sir William Hamilton and Sir Charles Bell, the eminent anatomist, had both formed conceptions of a sense which they denominated the "muscular sense." I find the most prominent sign of this sense shown in the sign for Weight. It will be found large in those who have exercised balancing of the body, in rope-dancing, climbing, billiard-playing, carpentering, and also in those who have been accustomed to estimate weight by handling or lifting. The great astronomers all exhibit this sign, and found this faculty and function most useful to them in their vocations.

THE HANDS AND FEET.

The hands and feet are exponents of the mechanical and artistic powers mainly, although very much more of the character can be gained from a close scrutiny of them. Whenever the individual possesses a large endowment of Weight, Form, Size, Order, and Calculation, the hands and feet will show the mechanical power predominant by an excess of bone, with a due admixture of muscle. Where the Brain and Nerve system prevails, the hands and feet will be small and wanting in mechanical shape and power. The artistic hand has more muscle and fatty tissue than the mechanical hand. Thus to a good observer the hands and feet are exponents of mental and mechanical abilities.

The art of chiromancy, once so popular, is based on Comparative Anatomy. Observers compared the hands of persons who were artistic, scientific, mechanical, or sensual, sufficiently to generalize and classify them. In looking over the writings of Desbarolles, Balzac, and others on this subject, I find that in the main their ideas agree with scientific physiognomy in regard to the meanings of characteristics of the hand. This branch of physiognomy is a very pretty study, and one can, by a little observation and practice, read character from the hand well enough to make a fine game for an evening's entertainment. The hand, as I understand it, corresponds in its method of growth with the order observed in the evolution of the rest of the human organism. The thumb, which is peculiarly a human member (no animal possessing a perfect thumb), reveals this order most admirably. The first or upper joint represents the Brain and Nerve system, the middle joint the Architectural or middle portion of the body, while the lowest joint, that connected with the hand, tells us of the Vegetative division; and these three parts of the thumb will in all cases correspond with the mental, artistic or mechanical, and sensual character of its possessor. As I have elsewhere treated of the hands and feet, further consideration of them here will be unnecessary.

SUMMARY TO RATIONALE OF FUNCTIONS.

CHEMICAL DIVISION.

In the preceding *resume* of the facial signs and meanings of the functions of the several organ-systems of the human organism, we find that the Chemical or underlying division of the human system and the lowest division of the face present nearly every characteristic and sign that is observed in the primary stages of primeval organisms. The functions here exhibited are those of assimilation or digestion, of excretion and secretion as exhibited in the kidney and intestinal systems, and of reproduction and respiration. The *faculties* evolved by the chemical action of these functions are the same that are common to animals possessing similar functions. Some naturalists class plants in this division, inasmuch as plants have the properties of sustentation, reproduction, and of excretion and inhalation in a certain sense. They take in through their numerous stomata one portion of the atmosphere, and exhale or throw off another. This is considered analogous to respiration, although the organs of respiration are not present as in animals and man. Yet all these vegetative functions are performed, evidencing a unity of action which is more observable the more profoundly we investigate Nature.

Other naturalists go so far as to offer proofs of will and sensibility in certain plants, notably the carnivorous class. Whether their claims are well founded or not, we cannot deny to plants chemical action and enough of fixed formation by means of their fibres to give them a place in the Architectural division of Nature; but I think when sensibility is demanded for them it may be disallowed. Notwithstanding my love and veneration for trees, plants, and flowers amount almost to a worship of them, I shall not let either my loves or beliefs conflict with scientific statements if I can prevent it. I do not approve of such mixtures, as they tend to mislead and confuse the mind.

The possession of *all* the attributes of mind cannot be denied the higher animals, for *facts* attest that many of them are superior in every sense, as regards reason, memory, intuition, invention, artistic and mechanical ability, calculation, honor, cleanliness, and morality, to the lowest grades of human beings, and they are in some traits superior to the highest. The faculties derived from the Chemical division are purely Vegetative. They exhibit dullness of intellect, sensuality, selfishness (of a negative sort), sluggish circulation, with an excess of fluidity; a great desire for assimilation, particularly of fluids, added to a love of domestic enjoyments; slow motions and inclined to inertia. The feelings are, owing to the softness of the constituents of this system, not positive; even love of offspring, which is well marked in this division, shows its power more in feeding and playing with children than in any way which requires activity or motion.

ARCHITECTURAL DIVISION.

This division exhibits all those functions and faculties which are produced mainly by the action of the mechanical or building powers inherent in the mechanism thus producing them. Within this department of the body are found illustrations of nearly all the principles of natural forces, and in the middle portion of the face the signs for all these functions and faculties are placed. The laws of acoustics, optics, pneumatics, hydrostatics, hydraulics, gravity, friction, capillary attraction, the lever powers, the pulley, the valve, different sorts of hinges and joints, together with galvanism and magnetism, find in this department their best illustration. The functions and faculties depending upon these principles for their power are the most varied in the entire organism. The functions which are exhibited by these powers are concerned in the form, size, motion, and action of the body; the faculties emanating from these functions contribute to the form and activity of everything which man does in the nature of architecture, and here I use this word in its most comprehensive sense. It includes the mechanical construction or building of all literary works, of sculpture, drawing, painting, designing, of dramatic representation, of the mechanical portions of music—of mechanical operations generally; in short, of every effort of man which brings into action any mechanical principle whatever. The more a man has of these principles in his own organism, the better he can express them in works. The functions in this division are performed by the muscles, bones, lungs, heart, liver, eyes, ears, hands, and feet. An examination of the principles which are involved in the action of these functions will show their capacity in the direction of architectural effort. An excess of action of the lungs and heart gives fine arterial circulation. This causes ambition, elevated sentiments, moral inclinations (observe how much morality of conduct is dependent upon fresh air and plenty of it), love and capacity for leadership, keen sensations, active sympathies, and noble aspirations. Those in whom the Muscular system is regnant are highly magnetic; hence, we must infer that the metallic substances which create magnetism are constituents of this tissue. I am not sufficiently familiar with histology to demonstrate this experimentally, but must content myself with the fact as illustrated in Nature, leaving the demonstration to those more capable who may chance to note this observation and statement of natural law.

MATHEMATICAL DIVISION.

The Mathematical division, or Brain or Nerve system, endows its possessor with great nervous and mental energy, activity, intuition, and capacity for receiving sensations, thus assisting by its intuitive perceptions and extreme sensitiveness in protecting the entire organism. This is its office primarily. This system in excess manifests reason, judgment, opinion, and mathematical ability. It also exhibits great electric power, as the Muscular system does magnetism. Electricity abounds also in animal organisms, notably in the torpedo fish, the Surinam eel, the *silurus electricus*, and several other species of fish, which are capable of giving electric shocks. This is their manner of defending themselves, and is probably a compensation for the lack of some other quality which would serve the same purpose. The Brain and Nerve system being the most delicate, sensitive, and the finest of all the organ-systems, is protected and sustained by a greater ,proportion of electricity than the other systems. This is the finest and most subtile force known to man.

The Vegetative division is sustained by its superior powers of assimilation and nutrition, and the Architectural by the strength of the mechanical powers, by the bones, muscles, heart, liver, and lungs. The Brain and Nerve system, being the highest and most perfected of all the systems in the body, is sustained by the superior quality of electricity evolved from its nervous and cerebral structures; hence its susceptibility to sensations, and activity, rapidity, and continuity of intellectual processes.

As in the body there are five superior organ-systems—five chief systems of functions and faculties—so in the brain there are five divisions, designated by naturalists as the fore-brain, twixt-brain, mid-brain, hind-brain, and after-brain. The fore-brain, which is popularly regarded as the portion of the brain where the intellect is located, spreads in its development or successive growths (as witnessed in the lower animal organisms and in the human embryo) nearly over the entire brain. Its three coverings—the dura mater, the pia mater, and the arachnoid, seem *diffused*, as it were, through the other parts of the brain, which they also inclose. Mr. Haeckel observes of the fore-brain:

",The highest activities of the animal body, the wonderful manifestations of consciousness, the complex phenomena of the activities of thought, have their seat in the fore-brain. It is possible to remove the great hemispheres of a mammal piece by piece without killing the animal, thus proving that the higher mental activities, consciousness and thought, conscious volition and sensation, may be destroyed one by one, and finally entirely annihilated. If the animal thus treated is artificially fed, it may be kept alive for a long time; for the nourishment of the entire body, digestion and respiration, the circulation of the blood, secretion—in short, the vegetative functions—are in no way destroyed by this destruction of the most important mental organs. Conscious sensation and voluntary motion, the capacity for thought and the combination of the various higher mental activities, have alone been lost.

"This fore-brain, this source of all these wonderful nervous activities, reaches that high degree of perfection only in the higher placental animals (*Placentalia*), a fact which explains very clearly why the higher mammals so far excel the lower in intellectual capacity."

Let me here digress a moment to call the attention of the reader to the last paragraph, as affording additional evidence of one of my positions; viz., that the perfection of the reproductive system and the perfection of *creative art* or talent in man are always found in combination. Indeed, all human character depends upon the fact of being *well sexed* for its perfection, for its energy, ambition, and general power in any direction. Power, Talent, Intellect are all questions dependent entirely upon physiology and anatomy.

The coincidence of five separate brain-bladders, as they are termed, evolving in the order of progressive evolution just as did the five organ-systems—viz., the Vegetative, the Thoracic, the Muscular, the Osseous, and Nerve systems, or the medulla or plexuses, as they evolved before any brain proper appeared—is remarkable and worthy of our investigation. In the absence of any positive proof on the part of anatomists and naturalists as to the separate and individual functional action of these five well defined growths or divisions, only theories can be presented. As in the body we find every function localized (or, if diffused, as are the muscles, glands, or some other organs, they can be traced by similarity of structure), so we must infer that each of these functions has its representative in the brain, or in some cerebral structure, and this includes the medulla oblongata. My theory in regard to localizing the consciousness of the several faculties and functions in the brain is this (and I argue entirely from the order of evolution observed in animal development and human embryology): That as the outer skin-covering was the first portion of the nervous system to evolve, it comprised the first sense-organ, or mental power; from this outer skin, by differentiation the nervous system and brain proper were evolved. The first organ-systems and functions which appeared were, first, the intestinal, and, second, the kidney system. Afterward all the other organ and sense-systems appeared by differentiation in the order previously stated, until the entire complex human system was reached. All these growths were the product of countless eons of geologic periods, each successive growth being ushered in by geologic conditions suited to its operations and necessitated by upward coeval development. Now, we know that there was a beginning to organs and organisms as such, and naturalists teach us that there were successive growths and superadditions of physical functions which accompanied or appeared simultaneously with so-called mental faculties;

that these functions and faculties appeared in the following order—that is, observing the order of the five superior organ-systems of the human body—1st, the Vegetative, or Chemical; 2d, the Thoracic; 3d, the Muscular, or Animal; 4th, the Osseous, or Mechanical; 5th, the Brain and Nerve, or Cerebrum. Now, if these systems evolved in this order (and we have proof incontestable that they did, and in human embryology we have the strongest evidence), we must infer that the intelligence or consciousness needed to operate these organ-systems was *evolved or developed in precisely the same order.*

I must believe that the way to localize the seat of consciousness or intelligence of these five systems in the brain (for I admit that all these systems have representation there as well as in the ganglionic masses and their attachments situated in the different parts of the body) is to trace the evolution of the brain and nervous system in the order of its successive growths or development. We know that in the invertebrate animals, those which have no brain, the mental and mechanical operations are carried on by means of the nervous system and nervous plexuses located in their bodies. The wonderful organization of communities of bees, for example, that form monarchical governments, divided into several orders, such as the queen bee, the drones, the nobility, the workers, nurses, and soldiers. All of these different orders perform their several duties according to their rank as methodically as do similar orders in a human monarchy. The ants, also without brains, form, on the contrary, a perfect republican system of government, managed as intelligently as the former, and, like our own republic in its early days, have the "divine institution" of slavery. They have slave-stealers and slave-holders; one species of red ants steal the small black ants and rear them to work, while others are set to tending plant-lice, whose milk or juice they are trained to collect. Thus man has unconsciously imitated the systems of government and industry, of cruelty and injustice, of these minute animals. This teaches us the lesson that creatures without brain, reason, or conscience, perform the precise acts of government, of wrong, outrage, and injustice, that creatures do who are believed to possess these fine qualities. With much truth, the monkey observed that "man is an imitative animal." Now, these creatures without brains do exactly as man does with brains. This makes good my position that intelligence in the lower animals is located in the bodily organs, without the necessity of one central organ or brain, as we find in the vertebrate classes.

In developing in a more complex manner, man *has not parted with the power for mental action derived from the organs and nervous plexuses* of the body, but has added thereto the superior intelligence exhibited by that portion of the brain where abstract reason is located, or brought into consciousness "in some mysterious manner," as Lewes remarks. Evolution has also enhanced the quality of the bodily functions and mental faculties—made them more complex, more diffused, and more intricately related.

I would here suggest to the anatomist who wishes to immortalize his name that he undertake to trace, if possible, the course of connection between the organs and functions of the viscera and senses and the cerebral structure, taking the evolution of the five organ-systems for his basis of investigation. He would do what no one has as yet been able to demonstrate scientifically and beyond all doubt.

For many centuries the Aristotelian theory of the circle held possession of the mind of the scientific world. It was argued that as the circle was the most perfect of forms, it must hence represent the orbit or path of the celestial bodies through space. Kepler proved this to be an error, and from that moment astronomy advanced with rapid strides. This idea held captive the minds of men, and impeded for ages the advance of truth. The theory that the brain is the *sole* and *exclusive* seat of mind, intelligence, and mental sensations has kept back for years the knowledge of the true nature of man—therefore of *true religion*. If we desire to progress in knowledge of the truth of God's laws, of scientific and exact law, we must utterly repudiate and cast out such monstrous error, and henceforth regard the *entire* organism of man as the *seat* of his mind.

I will close this summary by stating my belief that no one will rise from the perusal of these pages without, in some degree, modifying preconceived ideas as to the *rationale* of mental operations and the origin and locality of the mind.

The following exhibit shows the various organs and functions from which the several mental faculties derive their powers:

Conscientiousness... Kidneys.
Firmness............Bony System.
Alimentiveness..... Intestines.
Benevolence........Glandular system.
Amativeness........Reproductive system.
Love of Young .....Glandular system.
Mirthfulness........Glandular system.
Self-esteem ....Bony system.
Modesty.............Skin and Nerves.
Approbativeness ....Glands.
Friendship ....Intestines.
Resistance..........Muscular system.
Force.. ....Muscular system.
Secretiveness ....Muscular and Glandular.
Cautiousness........Muscular and Glandular.
Hope ....Liver.
Analysis.............Liver.
Human Nature......Healthy equilibrium of mind and body.
Imitation...........Muscular and Nerves of Sense.
Sublimity ....Equilibrated and powerful condition of the body.
Ideality ....Brain and Nerves.
Acquisitiveness .....Visceral vigor and Muscular development.

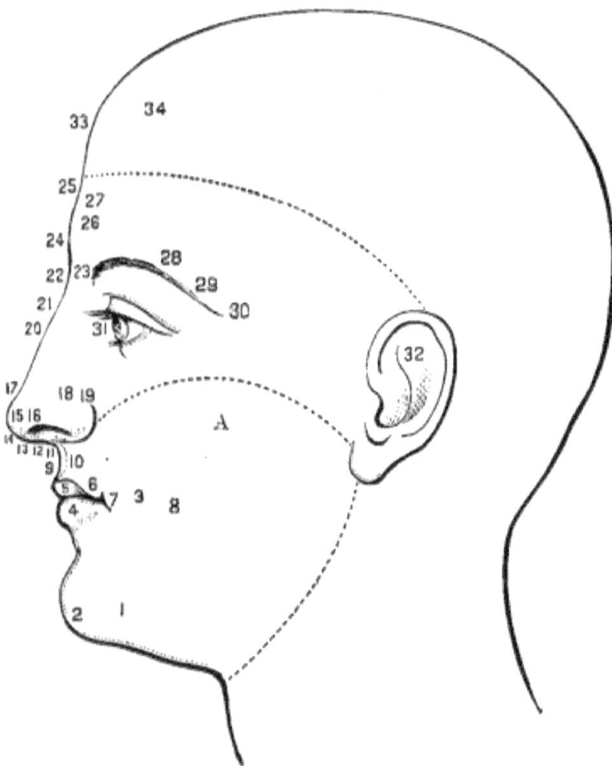

LOCAL SIGNS FOR THE MENTAL FACULTIES.

The above outline indicates by figures the positions of the local signs for the mental faculties. Many of them (particularly those about the tip of the nose) lie so closely together that keen discrimination will be required to locate them. The face being itself a microcosm, or miniature world, the difficulties of placing so many signs in so small a space, and have them clearly defined, must be apparent to the reader. For the sake of progress, it is essential that these localized signs should be comprehended thoroughly. I have not numbered such of the faculties as are general or that have more than one local sign; as, for example, Secretiveness, Resistance, Force, Form, Size, etc. These numbers are placed as correctly as can be, away from the living subject.

1—Conscientiousness.	12—Analysis.	24—Observation.
2—Firmness.	13—Imitation.	25—Memory of Events.
3—Alimentiveness.	14—Cautiousness.	26—Weight.
4—Benevolence.	15—Ideality.	27—Locality.
5—Amativeness.	16—Sublimity.	28—Color.
6—Love of Young.	17—Human Nature.	29—Order.
7—Mirthfulness.	18—Constructiveness.	30—Calculation.
A—Friendship.	19—Acquisitiveness.	31—Language.
8—Approbativeness.	20—Veneration.	32—Music.
9—Modesty.	21—Executiveness.	33—Comparison.
10—Self-esteem.	22—Self-will.	34—Causality.
11—Hope.	23—Credenciveness.	

Constructiveness.... Muscular system.
Veneration .... Bony system.
Executiveness....... Bony and Muscular systems.
Self-will........... Muscular system.
Observation .... Muscles and Nerves.
Form .... Bony structure.
Size .... Bony structure.
Weight............. Muscular system.
Locality........... Muscular system.
Color .... Glands and arterial circulation.
Order.............. Bony system.
Calculation........ Bony and Muscular systems.
Memory of Events .. Muscular and Brain systems.
Language........... Muscular system.
Time. .... Bony, Muscular, and Brain systems.
Causality .... Brain.
Comparison .... Brain.
Intuition .... Brain and Nerves.

CHAPTER VIII.

BEAUTY AND UTILITY OF THE GROUPING OF FACIAL SIGNS.

There are many faculties that do not develop to any, or scarcely any, appreciable extent until the age of puberty is passed. Some of the features will be found correspondingly wanting in character until the fifteenth year at least. The nose does not assume its just and true proportions until about that age. The forehead, also, does not until then assume a fixed form, and this, too, will change considerably in the succeeding twenty years if the brain be much occupied.

Faculties which require long use for their effect to be seen do not make their facial impress until late in life. Veneration, Causality, and Comparison do not show in childhood greatly, as these faculties not having been used long their permanent effect upon the muscles and bones has not been made noticeable, and the bones of the nose do not harden and become developed early. Still, the *forms* of the several systems of the body will point to what is latent, but especially will all the faculties in the organism be known to exist by the peculiarity of grouping, which is yet another method that Nature takes to stamp her meanings upon the face, and thus inform all beholders of the powers and peculiarities of the individuals so impressed.

Several of the functions of the body are not in operation in childhood, and their meanings not registered fully in the countenance at so youthful a period. Veneration is one of these; Executiveness still another; Self-will has not assumed its greatest expression; and thus it is that the ridge of the nose, where these signs are distinctly recognizable in adults, has not its full significance in youth. Still, the germ of all these faculties (where they form part of the organism) is there, and can be understood by the other signs in near proximity to them, so harmonious is Nature in all her works.

The signs of the faculties in the first division show out in early life more markedly than any others, for as these relate to the nourishment and sustentation of the body more particularly, they are dominant in childhood. Those relating to mechanism, art, religion, literature, and pure reason advance with the unfolding powers of the body. Still, the combinations of the several systems of the body will show by their conformation what will be the ultimate powers of the individual.

Let us commence our examination of the grouping with the first, or Chemical division. This includes those faculties which Phrenology has designated as animal propensities. The term animal, as applied by Phrenology to the part of the human organism which comprises the functions of nutrition, secretion, respiration, and reproduction, is a misnomer. The term animal applies properly only to such parts of the system as are endowed with motion, volition, and sensation. These functions cannot exist without some sort of muscles and nerves. These may be in the simplest form, or highly perfected, as in the higher species of organisms; but until such mechanism is supplied to an organism it cannot be designated "animal." These four functions are common alike to vegetable and animal growths, and therefore are recognized in this system of Physiognomy, as well as by scientists generally, as Vegetative.

I think the term "animal" is misleading, and tends to degrade these faculties in the estimation of the reader, inasmuch as they form the basis of the mind and moral character, and as it is owing entirely to the use we make of these faculties whether we become either healthful or moral. I believe it should be taught that they hold the most important place in the organism. The term Chemical is far more appropriate and expressive. As the operations of the faculties in this division are purely chemical, I can see no reason for naming them differently.

In the bony width and length of the chin will be found the most prominent signs of Conscientiousness and Firmness. Other signs there are for these faculties, which have already been explained. These faculties form the basis of character and commence in the physical powers, where justice and decision should predominate.

In the lower cheek will be found the sign of digestion. About the mouth and cheeks are other signs of digestion, Resistance, as well as the domestic endowments, Love of the Opposite Sex, Love of Young, Mirthfulness, Approbativeness, Friendship, Benevolence, Self-esteem, and Modesty. The beauty and harmony of this group must impress themselves upon all minds. These faculties, which are naturally and necessarily related to each other, here together find their most decided register. I have explained elsewhere, as the reader will observe, how the general signs may be located.

The next group, as we come higher up in the scale of developed character, shows us, first, the building powers; for,

after the sustentation and procreation of the race, the need
for shelter becomes of the greatest importance. Hence it is,
that in this division we find all that relates to the ability for
architecture, in its primary and broadest sense, and also all
that goes to show the qualities which make man himself
more perfectly built, and which exhibit his capacity in the
material world, for constructing, inventing, fashioning, shaping, and coloring ideas, images, buildings, works of art,
science, and mechanism.

Examine the group at the end of the nose. At the extreme
point, drooping downward, Imitation, the first essential in
building, in following and imitating the natural laws observed in Nature's processes, the principles of which are
found in man's own organization. Next above is Human
Nature; on either side, Ideality, or love of the beautiful;
the nearest sign to this, Sublimity; the next sign to Sublimity, Constructiveness; still farther back of the last, Acquisitiveness. The adaptability and harmony of this group are
quite apparent. As before mentioned, all the faculties were
primarily intended for the preservation, perpetuation, and
perfecting of the race. The placing of these signs is especially noteworthy as indicating the furthering of the perfection of man.

These local signs have more than one significance. They
not only indicate man's abilities in building from the materials of the physical world, but show by their location and
degree of power of each of these faculties within the organism of the possessor his ability to *transmit* and *reproduce*
them again in the human species. They show to us yet more.
They prove man's innate and instinctive perception and appreciation of these traits in others; for it is a well established law that one cannot fully comprehend in another that
which is greatly lacking in himself. Hence, he who possesses
these faculties is able to select in a companion those qualities
which in combination with his own will go still farther toward
the upbuilding and perfecting of the species.

Love of the beautiful is located on each side of Human
15

Nature, thus enabling him who possesses these to judge of the same in others. These two, with Constructiveness, which lies close by Ideality, give the power to construct as well as to perpetuate like faculties in offspring and in artistic creations.

Acquisitiveness, lying close upon Constructiveness, shows the desire and ability to acquire materials, powers, and resources for carrying forward the work of construction.

Analysis gives the ability to classify and suggest ways and means for the disposition of these materials, to advise in the construction, and lead by suggestion to the invention of necessary improvements.

These faculties, in conjunction with Hope, give all that power which is needed for selecting and constructing the human organism on a more perfected plan. Finally, and in their ultimate use, they enable us to create works of art, such as statuary, pictures, and poems, teeming with beauty and grandeur, breathing of hope and sublimity, elevating the individual to grander heights of imagination and more marvelous achievements in art. This grouping is not separated by any arbitrary sign of division, but each arrangement of faculties seems to be placed so as to give assistance to others related naturally to it.

Leading up from this last group, higher still upon the ridge of the nose, we find placed next above Human Nature the sign for Veneration. This faculty leads to respect for all things that deserve respect—God, law, righteousness, and old age. It yields ready obedience to what it considers right, and through the same feeling has power to assist in carrying out laws, regulations, and customs in consonance with the spirit of respect and veneration.

Next above this you will see the sign for Executiveness. This faculty is akin to Veneration in its nature; but, going still farther in the work of Veneration, commands by force, if needs be, the respect and right due. This faculty causes the highest outline on the nose.

The next sign, Self-will, is located in a position to give

aid and support to the faculties which cluster about it. Placed at the junction, as it were, of the artistic, mechanical, and religious groups, Form and Size are on either side of it; Observation, Memory of Events, Locality, Weight, Color, Order, and Calculation, in its immediate neighborhood. A decided Self-will, in conjunction with a good combination of faculties, either moral, religious, artistic, or mechanical, is a grand possession, as it affords the stimulus necessary for success. But, without a strong sense of justice, it leads to inordinate selfishness, ignoring the rights and wishes of others, and seeking only its own gratification at all hazards. King Henry VIII. of England was one of the latter class; the sign is large in his physiognomy.

The next group clusters about the eyes, and depends chiefly upon those organs for the power to do the practical work which Size, Form, Locality, Observation, Color, Order, and Calculation are called upon to perform. The more prominent are these signs in the face, the more practical the individual will be. These signs are all related either to the Muscular or Osseous system; they depend upon the predominance of those two systems for the power to carry forward the operations of the faculties of which these signs are representative. Some of these faculties are very nearly related; as, for example, Order and Calculation; and these two assist Music and Time. The most of these faculties which are found in connection with a large protuberance of bone, forming the bony ridge covered by the eyebrows, are related to the hands and feet; that is to say, that where these signs are very decided, the Bony and Muscular systems are in the ascendancy. Hence, the hands and feet are the channels through which these systems operate. Where the Muscular system dominates, the bony ridge is not so prominent, and the eyebrows are raised higher up than we find them in mechanical and practical individuals.

The beauty and harmony of the next group of signs must be apparent to the reader at first glance. Here pure abstract reason holds its sway, dominating the entire man; computing, by its mathematical powers, the combined resources of the organism over which it presides. Here intuition, time, rhythm, harmony, space, and dynamics are judged; the cause of all things is traced to its origin, and Comparison takes mental note of the differentiation produced by the complex action of the laws of Nature. It is in this department that man's superiority over the brute creation is more distinctly manifested than in any other. In the animal kingdom, the beaver and spider may teach man construction; the ant, mechanics and government; the elephant, sagacity; the fox, craft; the ape, imitation; the dog and horse, fidelity and patience; all mammals, love of offspring; but man alone is capable of abstract reasoning in its highest power.

The form and position of the local signs of these faculties in the face is an added proof of the truths of Physiognomy. Very much can be learned of this science by the study of the forms of animals and the characteristics which accompany them. It is in this department of natural history that the best tests of Physiognomy are found. Here all is openness and candor; no dissimulation, no "keeping up of appearances" from any motive whatever, but frankness, truth, and *natural* qualities are found; and Nature, not hampered or distorted by injurious habits, is seen in its true light. The forms of the animal world are left unfettered by appliances which would tend to deform or dwarf them, and the faculties, consequently, are in normal action. Here, then, is the best place for a commencement of the study of Physiognomy, as well as for its verification; for the forms and signs of faculties in the animal indicate the like faculties in man, and are accompanied by the same forms and signs. In form, color, quality of hair and skin, all have a similar meaning, and man needs only the degree of observation given to other studies to become the interpreter of Nature's most palpable and unfailing indicator of faculties. Hence it is, that in their traits and forms, animals foreshadow and typify like conditions in man; depending for their explanation on the same colors, forms, and textures, they will be found identical in both; therefore, a close observation and keen analysis of animal life, together with the study of recent discoveries as to their habits, will advance the student incalculably in the knowledge of human Physiognomy. A work entitled "Mind in the Lower Animals," by Lindsay, will be found most interesting and instructive to the reader.

CHAPTER IX.

FORM AS A UNIVERSAL AND DETERMINATING PRINCIPLE.

" Unity of form resides in every department of the universe, and binds the whole in one."— BABBITT.

Among the principles which influence and determine character throughout Nature, the Architectural must be ranked as foremost. The *form* or *shape* of everything throughout organic, or even inorganic, life testifies to its character and rank among creations. Every mineral has its own form of shaping. The law which causes minerals to crystallize always in certain shapes is not understood. The fact that they always do assume certain fixed architectural or geometrical proportions is patent to all who have made the slightest investigations of mineralogy, and one accustomed to see the various forms in which the mineral constituents of the earth compose themselves to rest can tell, by the sense of feeling, even, without seeing them, the different crystallizations of minerals, by the number of sides, or facets, which each mineral assumes. Sulphur always affects two distinct formations; quartz fashions itself, or is fashioned by the law which governs it, into prisms; and, however it may vary in shape or size, it always presents the same number of facets— six. Feldspar shows ten facets; silica forms thin, flat leaves; snow always exhibits six facets, although it produces many beautiful diverse crystals. Salt, too, has its own form of crystallizing. After their forms and accompanying properties are understood, their forms thereafter indicate their character.

It is just the same in organic life. Every blade of grass, herb, weed, flower, shrub, and tree comes to be known by its form. The character of animals is determined in the same manner. The crouching tiger, eager for his prey; the timid hare, with its frightened look and long, thin, erect ears, announces each his character by his form. When looking at the great beasts of prey, we know what their forms indicate. If the mouth of an animal is wide and furnished with great fangs, we know that he is a great eater and carnivorous. In comparing the domestic animals—the camel, horse, and ox—with the former, we know, by their form and general appearance, that they are docile, as well as strong. When one observes the eagle, vulture, or falcon, with their clear, piercing eyes and arched beaks, he instinctively feels that these creatures are representatives of strength; and wherever we observe the arched form in Nature, we shall find superior power. Man, in his imitations of Nature's architectural designs, uses the arch wherever great strength is required. This imitation is one of the manifestations of the same law and power which fashioned the humps on the strong camel's back, and the beak of the bird of prey.

In the human as well as animal world, like principles illustrate similar characteristics. Straight form and lines denote straightness of action; curved or crooked lines indicate ability and desire to move in curved rather than straight lines; squareness of formation shows the ability to act "on the square;" and uprightness and downrightness of conformation reveals upright and downright characters. Circular persons love motion better than square-built ones, and are more capable of continued motion, especially in dancing, etc. Then, too, round persons and animals pursue more wandering lives than those who are square-built, each being true to the law of his conformation. The muscular races of people and animals travel faster, more, and enjoy it better, than those of the bony formation. Among animals who are given to excess of motion, besides the carnivora, I may mention the various deer tribes, such as the reindeer, gazelle, springbok; also, the greyhound, the hare, antelope, etc.; while the ox, elephant, camel, dromedary, dog, and horse are more inclined to slower motions, and enjoy a settled domestic existence. They are also more docile and teachable, and susceptible of more improvement, than the nomadic tribes.

In man's organization, the same appearances indicate similar powers. The high, square shoulders speak of strength; the drooping shoulders, of weakness; the arched chest, of great lung-power; the flat chest, of an enfeebled, weak condition of those organs. The prominent joints of the bones tell us of a powerful framework; the small joints, hidden by fat, announce a less useful organism. Prominent features disclose to us a strong, ambitious character and corresponding physique. The broad, dome-shaped head reveals a capacious brain, as well as a body stored with vital material. The dome shape in architecture always exhibits more strength than the flat or square form; it shows more power of resistance from every quarter. Hence, we find that round features of the face denote power of resistance; arched features, ability for overcoming. The high, arched

NOSES,

arched from the root to the point, and rounding at the sides, indicate commercial capacity, the same faculty which is possessed by the birds of prey—the faculty which profits by others' industries, without itself producing anything. This form is noticeable in those who have excelled in the world of commerce. Sir Moses Montefiore, the Rothschilds, Stephen Girard, John Jacob Astor, and others of like character possess this formation. Vanderbilt's nose is somewhat different from these; his nose is the executive combined with the commercial. Wellington possessed the executive nose in its most decided form; indeed, it is not unlike a battleaxe in shape; very superior in this respect to Napoleon's nose, who met in him his conqueror. Portraits of Julius Cæsar, Hannibal, Frederick the Great, Napoleon Bonaparte, General Scott, the Napiers, and many of our generals in the war of the Rebellion, were characterized by arched or commanding noses. Those possessing this form are in danger of becoming overbearing and tyrannical. It requires a good degree of benevolence to balance this formation. This nose, rising high at the "bridge," always indicates superior physiological construction, as compared with the low or "scooped" nose. The arched nose tells us of a strong stomach and lungs, and these argue other powers which I have referred to elsewhere.

The nose reveals the mental, literary, artistic, commercial, executive, and mechanical powers; hence the importance of having a good nose and "plenty of it." Small or flat noses indicate weak characters, with no power of command—not even of themselves. If the nostrils are so placed that one can see the interior from a front view, such character is selfish and passionate; such will manifest little power of self-control, but will live in their will, feelings, and emotions. Where the nose is long in proportion to the other features of the face, a cautious and thoughtful character will be its companion. A long, high, thin, bloodless nose is often indicative of consumptive tendencies; thin, narrow nostrils would, of course, denote narrow, weak lungs. A nose which shows a separation into two parts, or a cleft at the tip, exhibits the character of a natural critic. People with such noses will naturally criticise everything which attracts their attention. The *kind* of criticism is indicated by the combination of their faculties. If they possess the color sense largely, they will be able to analyze well all combinations of color; with the literary sense developed, they will criticize intelligently literary efforts; but they will be able to analyze and criticise naturally whatever falls under their observation.

Short noses, rather inclined to turn upward, belong to willful, reckless, and destructive persons—often impudent and saucy. This form of nose is seen sometimes in the physiognomies of prize-fighters. Portraits of Heenan, Tom Sayers, and Yankee Sullivan show them to have been possessed of such. Bony noses show firm, argumentative, and executive qualities. Muscular, cartilaginous, or fat noses disclose yielding and flexible qualities, such as are needed in the imitative, artistic, and decorative efforts of the mind; hence, when a nose possesses more of the bony than muscular development, the character is more remarkable for strong sense, firmness, and integrity than the nose which has more of muscle or fatty tissue. Here, as elsewhere, the muscles indicate power for feeling and expression, emotion and passion; for muscle is relied upon by art to express its operations. Short, round, full noses belong to artistic and musical people. Wilhelmj, the celebrated violinist, possesses this form; also, Madame Parepa Rosa, vocalist, and Arthur Sullivan. composer. Many painters have short noses, low in the centre, but broad and blunt at the tip. The signs of Ideality, Constructiveness, and Sublimity are manifested, but the executive are entirely wanting. Yet great and original artists have large Executiveness, Veneration, and Self-will added. All these signs are decidedly prominent in portraits of Salvator Rosa, Michael Angelo, Peter Paul Rubens, Vandyck, Da Vinci, Titian, and all others who have shown superior original works. Originality implies force and courage, and a small nose never exhibits these qualities in artistic works. Such, not having the power to originate, always follow in the beaten path, and are mere copyists.

Among dramatic artists, the same differences are observable. All of the most original and powerful actors and actresses possess noses large in Self-will. This is one of the most essential faculties to those actors who sustain leading *roles*. They must be able, by *force of will alone*, to hold the character which they are depicting through several long acts; and this requires something more than mere imitation, especially in grand characters, like Richelieu, Queen Elizabeth, Marie Stuart, Adrienne Lecouvreur, etc. Accordingly, we see in the portraits of those who have excelled in the highest walks of dramatic representation, noses with Self-will, Executiveness, Veneration, Sublimity, Ideality or Imagination, Constructiveness, Human Nature, and Mental Imitation very large, together with Self-esteem, Amativeness, and Language. Madame Ristori, Sara Bernhardt, Talma, Edwin Booth, David Garrick, Sarah Siddons, Madamoiselle Mars, John P. Kemble, Junius Brutus Booth, and other first-class dramatic artists, have all of these signs in their physiognomies. These first-class actors I designate "creative actors." There is another and far larger class of actors, who resemble each other in features very closely, or have essentially similar features, faculties, and signs in the face. These are most noticeable in the shortness of the upper lip, fullness and size of the eyes, size of the end of the nose, and thickness and redness of the centre of the upper lip (sign for Amativeness), and the sign for Love of Children and pets. This large class can only be considered as "imitative artists." They are known by the hundred, all exhibiting the same characteristics. The upper part of the nose, where Self-will, Veneration, and Executiveness are located, is deficient, more or less, in most of this class; while Language, Amativeness, Mirthfulness, Love of Young, Imitation, Approbation, Imagination, and Hope predominate. These signs are disclosed by a short upper lip *invariably*, upper part of the nose not prominent, sometimes depressed, very broad at the point; sides of the nose full at mechanical ability (this faculty is needed to produce the mechanism of the characters represented), with a decided droop below the tip of the nose (Imitation), and a very perceptible projection downward of the septum at Hope and Analysis; the mouth at the corners turned upward (Mirthfulness); Love of Young, Human Nature, Secretiveness, and Amativeness well developed, forming the "Cupid's bow" shape, which is observed in nearly all artistic and literary persons; Amativeness always *well* developed—sometimes too much for either beauty or morality; yet this faculty is essential to the creation of the characters which they portray.

This description applies to hundreds of male and female representatives of dramatic art of the imitative class—to those who excel in burlesque, in character songs, in negro minstrelsy, and in what is called variety show business, and in all the minor parts of the mimic art. This single description will answer for all, although even in this class there are many grades of ability. Lydia Thompson, the Worrell sisters, Pauline Markham, and others of this type answer to this portraiture.

Very long, keen, sharp-pointed noses are illustrative of sharp, keen, ferret-like characters, without much imagination, very matter-of-fact, "high-pitched," and excitable. The noses of these persons are gimlet-like at the point, and seem the right size for penetrating a key-hole. Being unimaginative, they naturally look for matters of fact which may be transpiring on the other side of the door.

Persons possessed of elegant tastes, refinement, love of poetry, and capacity for belles-lettres, will in most cases indicate this taste by a rather prominent and straight nose; that is, straight in the outline from the root to the tip; the end may be broad, showing Ideality large, and the sign for Mental Imitation is often found with the straight nose. This nose to be effective must be of good size. The straight nose is seen in the portraits of Shakspeare, Byron, Moore, Milton, Beranger, Corneille, Swift, Addison, Burns, Lamb, Elizabeth Barrett Browning, Shelley, Mrs. Hemans, and many others who have excelled in the higher walks of elegant literature.

The large high and straight nose is seen on the faces of many eminent scientists, but the descent from the forehead at the root of the nose, where it joins the forehead, differs from the above mentioned class. The brows of scientists are more projecting, showing Weight, Observation, and Locality to be more defined. This is the case with Proctor, Tyndall, Helmholz, Huxley, Spencer, John and Joseph Le Conte, and Draper. Professor Edison's nose exhibits more of the mechanical and scientific, as well as inventive faculties. Professor Proctor has the sign for Weight very large;

he is also a *round, large, heavy* man, thus showing his capacity (through similarity of structure) to comprehend the nature of ponderous round bodies, like the planets, while Professor Edison has a slight, thin, nervous form, thus by his physique evidencing his ability to cope with and comprehend the principles underlying the *finer forces* of Nature. The electrical is more particularly the distinctive quality of the nervous and mental system predominant, while the muscular system of Proctor gives a better comprehension of motion; all of which accords with Dr. Simms's theories on this subject, which I indorse. Thus each of these men succeeds best in those branches of science whose principles are exemplified in a large degree in his own form and organism.

I have before me a chart on which are engraved the physiognomies of twenty-two prominent Jewish Rabbis. The similarity of their features is most striking. The noses, eyes, and mouths have all something in common. The noses of all are high and straight, and join the forehead without any indentation, showing that dogmatic Self-will which characterizes the advocates of all ancient faiths. They are not argumentative noses. This is not their *role;* they are not here to argue, but to assert. This same peculiarity is observable in the expounders of the Hindoo religions, as well as in most of the Catholic divines. The noses of all these Rabbis are very long, quite high, and wide at the nostrils, and showing by the width of the tip the signs of Sublimity and Ideality, and by the breadth of their nostrils the vigor of their visceral systems. The eyes of all are large, round, and prominent, denoting power of expression in language and emotion. The mouths of all disclose Amativeness, Love of Young, and oratoricel power, being straight and wide, thus revealing not only volume of sound, but good digestive capacity. Some of these traits are social; others are common alike to all who expound ancient dogmas as a profession. Love of young, strong appreciation of the opposite sex, mirthfulness, and good digestion are race peculiarities of the Jews, as they are of all the ancient civilized races. These all tend to make the character home-loving, domestic, and substantial. Their noses exhibit great Veneration, and this, with arching eyebrows (Credenciveness), gives a highly devotional cast of mind. All these Rabbis' noses are broad on the top their entire length. I should like to present my readers with charts devoted to the faces of distinct classes of people, but the shortness of my purse forbids.

This class of faces is just the reverse of the scientific or mechanical; the former resting its ideas upon faith and belief does not require that power of exact observation and comprehension that the latter does. The arched, rounding, or full shape of other features beside the nose reveals strength. Full cheeks show assimilative and digestive powers; flat or hollow cheeks the deficiency of this function, and often dyspeptic and consumptive tendencies. A thick full upper lip proves **Amativeness** and the power of the reproductive system; the flat thin white upper lip denotes a lack of both these important qualities. A full rolling lower lip, a sympathetic, benevolent trait; a thin flat white under lip, a miserly, unsympathetic person. High cheek-bones indicate self-protective capacity; flat cheek-bones or those not so prominent, less of this power. The rounding form of the arms, legs, and hips, as well as jaws, cheeks, and lips, indicate more strength than those which are opposite in form.

EYES.

The convex or rounded eye evinces strength. The eye of magnitude is capable of receiving more powerful impressions than the small eye; hence it is that the eyes of painters are large and wide, capable of taking in at a glance immense scenes of Nature. The eyes of all the great artists of the world are the proofs of these statements. Small eyes, on the contrary, belong to the mechanical, scientific, and exact people. The mechanic and scientist have need of the greatest accuracy in detail; their knowledge sometimes depends upon the hundredth part of an inch, and the scientist deals in atomic proportions. The eye which can be most readily brought down to a focus serves this class of minds the best. The eyebrows of mechanics and all keen observers are drawn forward and downward, to assist the eye in its scrutiny of minute objects, as well as to shade it. The form of the Bony structure assists this intention, for in natural mechanics and scientists the lower part of the forehead projects, thus affording a natural screen to the eyes, obviating the use of the glass which one sees the watchmaker and naturalist sometimes employ to focus their vision, and scan infinitesimal objects.

One proof of the eye being the exponent of the Muscular system, as laid down in the chapter on "The Rationale of the Physical Functions and their Signs in the Face," is found by observing all those animals which depend mainly upon their muscles for their activity. The eyes of the various kinds of deer, the springbok, gazelle, goat, ibex, elk, chamois, hare, and rabbit, all indicate by their size and conformation that the Muscular is their predominating system. It also exhibits their capacity for continuous and rapid motion. The size of the eyes, as well as the habits and habitats of these animals, prove that they are intended to scan distances and wide expanses, and that they are not suited to the perception of the minute in Nature. Small-eyed animals exhibit faculties just the reverse of large-eyed ones, and depend more upon the knowledge of things near them and those that require more exact vision. These animals are slower in their motions, possessing more bone and adipose tissue than muscle according to their size. The elephant, tapir, rhinoceros, grizzly bear, walrus, and elephant-seal are examples of this combination. The small eye of the elephant is so easily brought to a focus that he can pick up a needle with his proboscis. The nature and habits of all these creatures demand that they shall take cognizance of small objects, as well for the purpose of gaining a supply of food as for protection, their motions being so slow that accurate and instant vision is essential to their safety.

The mental characteristics of animals possessing large, full eyes bear a strong resemblance to human beings who have similar features. They are more emotional than the small-eyed creatures, more affectionate, and more active; they receive sensations more vividly, and lose them almost as readily as they receive them. One class of animals in which the Muscular system predominates exhibits considerable mechanical, and even artistic skill. The mole's burrow and the beaver's dam are manifestations of one form of muscular ability. In man, all the varied architectural and building powers depend mainly upon the Muscular and Osseous systems combined.

In undertaking to discover the meaning of an eye, there are *seven* things, at least, to be observed: First, the form; second, the size; third, the color; fourth, the degree of brightness; fifth, the shape of the commissure, or fleshy opening, caused by the parting of the upper and under lids; sixth, the angle of inclination, or the manner in which the eye rests in its socket—its inclination forward from the lower part, and its position in regard to the surrounding parts; seventh, its general expression. These are some of the expressions which the physiognomist must observe if he reads the eye conscientiously. There are many expressions of the eye which cannot be described in any way apart from the living subject, just as there are many other indescribable traits in persons that cannot be illustrated by brush or pen. The personal atmosphere, or magnetism, is one of the things which cannot be transmitted to posterity by words or pictured representations.

The enthusiasm and inspiration of the ancient orators—Cicero and Demosthenes, for example—must have produced the most exalting and sublime effect upon their hearers; yet, in reading their noble and lofty sentiments, this is all lost. Just so with the human eye. Many of its expressions cannot be reproduced. The various colors of the eye are often difficult to delineate on canvas. The best study of the eye must be made in the living subject. Still, there are many expressions observed that can be described, and their characters indicated by their color.

As a general rule, dark or black eyes express capacity for the passions—Love, Jealousy, Revenge, and Hatred. Black-eyed people are good lovers and good haters. A great deal of pure passion is mingled with their love—and strong will, too. Yet, where such' are harmoniously mated, they will be faithful and devoted, and defend the interests and honor of the beloved one with much ardor and zeal.

The true blue eye exhibits amiability, calmness, and steadfastness of purpose, sentiment, affection, and capacity for improvement; seldom very jealous, and not revengeful; it also shows a truthful, confiding nature. Hazel eyes betoken vivacity, sprightliness, ability for plotting, planning, and scheming; often insincere, although having an innocent, infantile expression of the eye. Yellow eyes evince cruelty, insincerity, and great power for deception. This color is seen in the eyes of tigers, panthers, and cats. Other traits must be looked to for other characteristics. Many persons with this color of eyes are often very entertaining, with a soft, pleasing voice, and attract many. Green eyes, or green spotted with yellow, are evidential of excessive jealousy and suspicion. Mixed, mottled, or spotted eyes show a mixture of races—a crossing of two or more races within two or three generations. This crossing often produces talent, but at the same time induces jealousy and suspicion. It seems as if the opposing elements of the mixture of the blood and tissue had not fused or harmonized. After two or three generations, the eyes become uniform in color, and the character better balanced.

Eyes which are placed near together are deficient in the comprehension of form and size. I have known persons of superior education with this peculiarity, who, by reason of this defect, were unable to remember the forms of words, and were consequently poor spellers. I cannot think any person entirely frank and thoroughly well principled, who is narrow between the eyes. Such will always be politic and secretive, and secretiveness is for the purpose of hiding something. Let the observer find out what is hidden. Narrowness between the eyes is evidential of a narrow mind; no matter how brilliant the person may appear who exhibits this conformation in his physiognomy, he will not be able to take as broad and comprehensive a view as one very wide between the eyes. History shows no man of eminence with such peculiarity. In selecting a horse, the intelligent dealer always chooses one who is characterized by width between the eyes. The same rule is observed in choosing dogs for their intelligence; a dog wide between the eyes is very easily taught tricks; he remembers form and size well. Then, too, where the space between the eyes is great, the median part of the brain is wider than where the faculty of form is deficient.

Very light eyes denote weakness, generally of a scrofulous or consumptive tendency. This color is sometimes accompanied by weakness of vision and deafness, eruptions on the skin, and scrofulous swellings, as is often observed in albinoes. Persons with these indications should never intermarry, as the result to offspring would be highly disastrous. The mentality of light-eyed persons is never of a very gifted character; they are from the very conformation and quality of their organism weakly; their minds partake of this enfeebled condition; they are generally surface people, fond of dress and show, with little sympathy for others, and much absorbed in the adornment of self; can never become good colorists in the arts and industries. This type of people would soon die out, if intending parents would abide by physiognomical law and cease to perpetuate this almost helpless and useless species.

Gray eyes are often the accompaniment of talent of a calculative, mathematical, and philosophical order. I speak now of eyes of the color of milk and water mixed, with a decided bluish cast. Many persons denominate brown and hazel eyes gray; indeed, many have a confused idea as to what constitutes a gray eye. This bluish-white eye is cool

16

rather than ardent in love, shows good worldly planning ability, and is inclined to be suspicious. The absence of color in the organism is a proof of a great defect in some part other than the skin. This defect is not only physiological, but it naturally illustrates mental deficiency. The color sense is not usually strong in this type, yet with cultivation can do very well with colors. Persons with gray eyes, where the white preponderates over the blue, are often subject to diseases of the kidney system and pain in the small of the back. In females, the reproductive system as well is affected, and indicates mechanical obstruction in its functions. We cannot ignore the fact that a defect in the chemical combination of the materials composing the human organism induces diseased conditions of the several systems and functions, and as a consequence diminished mental activities and moral proclivity and power. Color is a chemical effect produced by a variety of causes. It emanates from the sun's rays, from the atmosphere, and is extracted from the food we eat. It is essential to our mental and moral welfare that the right proportions of color should be mingled in our food in the natural way, and that we should obtain from the sunlight sufficient of its color, chemically combined in our organisms, to produce moral, mental, and physical harmony, without which man's organism cannot be moral, healthy, or perfectly balanced.

The sign for voracious appetite, or gluttony, is found in the face of the person whose eye projects outward and forward at the lower part, disclosing a good deal of the white of the eye below the pupil while the eye is in its natural position. With this formation we may safely infer that the individual is given to excess in eating, and such are affected with headaches of long duration, the result of gorging themselves. The instant I perceive this eye I know that the individual is a victim to severe headaches. It requires no magic to diagnose this symptom. We all love to follow our proclivities, especially where they are very pronounced. This sign for appetite lies adjoining one sign of the intestinal system, and is an indication of an excess of its activity, or, in other words, of its abuse.

THE MOUTH.

In looking for signs of character in the face, one should pay close attention to the expression of the mouth, and observe what it discloses. The wide straight mouth, with lips that are evenly developed and full, or moderately so, is never found except in the visage of an eloquent person. Demosthenes, Cicero, Mirabeau, Clay, Webster, William Pitt, and all great actors and actresses possess these characteristics. A mouth, the lips of which are unequal, where one is much larger than the other, or where the under projects far beyond the upper, is indicative of an unbalanced mind and temper. Lavater tells us that "all disproportion between the upper and under lip is a sign of folly or wickedness."

The mouth which turns down at the corners is found with a sad, joyless, melancholy nature. This is true also of all features and lines which tend downward. If the nose turns toward the earth, there is more indication of melancholy than in the nose which points heavenward. Such noses always evince sanguine and hopeful dispositions.

Small mouths do not as a rule exhibit great talking propensities, but show refinement in the use of language, yet never eloquence. A small mouth, with large eyes, denotes a great talker, not necessarily a good talker. There are many grades in vocal expression, ranging from "gab" to conversation, and from conversation to eloquence. The mouth, ear, and eye will give the clue to these grades. A large mouth and full lips are commonly thought to indicate a gross, sensual person. It depends entirely upon the quality of the individual with these features whether he be such. A negro with these features very pronounced would undoubtedly be a coarse, sensual character, because his tone or quality is coarse; but no man of a fine social, benevolent, or domestic nature is found with very thin lips. An orator or a linguist could not be celebrated were he lacking in this particular. Full lips denote linguistic and social abilities, as well as domestic and sympathetic qualities.

The gums of the teeth are indices of character. Persons whose gums show greatly in talking or laughing are as a rule very commonplace people, and incline to some form of scrofula. I have never known a truly great or talented person who exhibited this peculiarity. Gums that are pale and bloodless denote an ænemic state of the body; they show an impoverished condition of the blood and weakness of the entire system. Where the gums are of a deep dark red color there is a diseased condition which should have hygienic treatment. Gums covered with tartar or a thick yellow accumulation are most revolting, and certainly contain the germs of serious disorders if allowed to remain long without treatment. The gums should present a healthy pink color, and be kept clean and of a normal degree of hardness.

HAIR.

In reading character, one must not pass by the meanings which the hair discloses. The *colors* of the hair illustrate essentially the same principles, mentally and physically, which the colors of the eyes reveal. Black hair is significant of passion, strength, warmth of feelings, and intense emotions, as compared with lighter hair. Very light or almost white hair is often indicative of a feeble constitution and a scrofulous diathesis, and is never accompanied by deep feeling. The reason for this is physiological; the entire organism lacks strength, both of transmitted quality and acquired vitality. Light-haired people are often, showy, sprightly, and amusing, but I have never seen a profound thinker in this class.

Of the various shades of yellow hair, ranging from molasses-candy color to flaxen, I can only say that they are not unlike the very light shades in their significations. These hues are generally found on the heads of persons more en-

tertaining than philosophic, whose emotions are transitory
and manners gay and lively; fond of dress and amusement,
and exhibit a great fondness for spectacular plays and bur-
lesques, and the sensational in literature; with the color
sense cultivated are very ingenious in many kinds of ornamental work. Such persons attract by their vivacity; their
affections are not deep, strong, or lasting, but fickle and ca-
pricious. That these shades of hair are not indicative of the
most developed characters we have only to refer to infants
and children of the Caucasian type; the hair of these children in most cases deepens in color as the body and mind
strengthens and develops. We must therefore conclude that
very light-haired persons are more infantile in their natures
—that is, not so matured as those who have deeper hues
in the hair. Yet black-haired persons are not so progressive
as brown-haired people. Dark brown hair, if not too fine,
evinces strong bodily and mental powers; if very fine, then
the mental predominates over the physical. All of the dark
brown shades of hair denote a good degree of intelligence,
with amiability, good sense, and deep feelings without bitterness.

Curly hair shows a changeable character, often brilliant,
vivacious, quick-tempered; sometimes possessed of imitative
talent, with a good deal of "snap" and "dash;" not very
steadfast, but desiring to change back and forth; sometimes
sunny and sometimes cloudy, like April weather, and with a
good deal of constitutional vigor, the circular form here, as
elsewhere in Nature, denoting superior strength.

Hair that lies in waves and rings is often seen on the heads
of gentle, amiable persons. Many talented people have this
peculiarity. Stiff and straight hair reveals a character very
firm and decided and inclined to honesty. Curly hair does
not show the same degree of integrity that stiff straight hair
does; yet many curly-haired persons act honestly. Conscientious action is induced by different motives; sometimes
from clear, natural, strong conscientiousness in the build or
make-up of the individual, sometimes from policy in busi-

ness and society, sometimes by the power of the affections, and at other times by necessity or compulsion. The affections influence many to pursue a straightforward course, in order to make provision for beloved objects; yet no form of conscientious conduct is as strong, genuine, and heroic as where it is the dominating element in the physique of the individual. For such it is *easier* to be honest and honorable than to be otherwise.

The light shades of brown hair which sometimes are found in combination with blue or gray eyes are generally indicative of good intellect, and evince physical and mental powers neither very weak nor remarkably strong. If the hair be fine it denotes delicacy of thought and feeling. Many persons of poetic ability have possessed light brown hair with a golden tinge. This combination betokens excitability and an exalted condition of mind, which often eventuates in expression by pen or voice.

Very coarse hair belongs to coarse, strong individuals, of a low grade of mental power, often rude, boisterous, and unsympathetic. The contrast between the coarse, strong, straight, stiff hair of the North American Indian, and the African's curly, woolly head, is as striking as are the differences in the characteristics of these two races. The best specimens of Indians, especially the northern and eastern tribes, show as much Conscientiousness in their faces and physiques as can be found in the most cultivated races. This trait the Indians always exhibited in their lives, until demoralized by men professing Christianity. They are not, however, very marked for kindness or sympathy. The negro, on the contrary, is a "curly character," with very little honesty in his composition, with little firmness or heroism; unreliable, yet sympathetic, generous, and sociable; with strong natural affection for offspring, with great Amativeness, yet not stable in his attachments, being a natural polygamist, as evidenced by the shape of the commissure of the eyes, which are almond-shaped; that is, longer from side to side than from top to bottom.

Red hair is evidential of ambition, ardor, and natural delicacy (where it is fine). The skin of red-haired people is generally very fine and clear. I infer that there must be more sulphur in their composition than in others. Now, whenever I observe a clear skin, I naturally look for clearness of intellect and moral inclinations. Red hair shows quick temper, lively and intense emotions, great agreeability, amativeness, and a love of out-door life and active pursuits. As the texture of the hair and skin discloses the quality of the Nervous system—therefore, of the mental power—we must infer a strong relationship or connection between them. The evolution of man proves that the outer skin-covering, or exoderm, in the primitive organisms assisted in forming the Nervous system. This discovery in the history of the lower organisms teaches us how this relationship came about, and proves that the skin and hair, eyelashes, and finger and toe nails were all evolved from the outer skin-covering. This knowledge gives the only solution as to the cause of gray hair. The hair of persons who have received great and sudden nervous shocks has turned gray or white in a short time, and sometimes in a single night. This is said to have been the case with Marie Antoinette when she was imprisoned. The hair does not generally commence to turn gray until the nervous power has begun (in common with other physical powers) to decline. Sometimes the hair turns gray prematurely in youth. This is often an inherited peculiarity. For this, as well as for gray hair produced by age, I know of but one safe application: A dozen or more of nails steeped in black tea, and the decoction used on the hair daily, will keep and restore the hair to its natural color. In most cases, it will prove a perfect remedy.

Very long, strong, and luxuriant growths of hair and beard betoken strength, longevity, great reproductive powers, and descent from a long-lived ancestry. Thin, scattered, fine hair denotes delicacy of constitution, fine and keen perception, sensitive and often shy natures, with nervous irritabilty, and sometimes brilliant mental powers; although other signs must discover the latter—no *single sign* will give the entire character of any one.

All these signs of character shown by the hair are to be considered with discretion and judgment. Without discrimination, all signs fail. The mole, hare, and rabbit have fine, thin, short coats of hair, and are sensitive, shy, and short-lived. Soft, pliable hair is evidential of tractable, amiable dispositions, not as rigid and "set" in principle, but more easily swayed by emotion and affection, than stiff, straight-haired people.

THE EYEBROWS.

The eyebrows and eyelashes are judged by the same rules that govern the hair, both as regards their color and texture. Thin and fine eyelashes and eyebrows denote sensitiveness and delicacy of the Nervous system and constitution generally; also, mental ability; that is to say, the mind will be either keen and sensitive, or profound and philosophic, with this appearance. Thick lashes and bushy eyebrows show strength and vigor of constitution; and, if dark or black, strong passions as well.

Among the people of all the civilized races, one meets with singular combinations in the colors of the hair and eyes. These combinations, being opposite in color, would seem to indicate traits not in harmony, but an analysis of those who possess them will show that in this, as in all her works, Nature is harmonious. The combination of blue eyes and black hair forms a very beautiful contrast, and one might infer that this indicates violent contrasts in the character, but it is not so. There seems, in this case, to be a fusion or blending of faculties peculiar to these two opposite characteristics. Persons of this type have great control of their feelings, with ability for planning, plotting, and conspiracy, and can carry out their plans in a very secret manner, while at the same time affecting an air of frankness. The combination of brown or black eyes with fair hair is not as common

as the former. Its meaning is strength and fineness; some
of the faculties will be very decided, while others will be
weak. This is a combination well worthy of study; its contrasts are more striking than those of the former.

DIMPLES.

How are dimples caused, and what do they mean? Dimples are caused by, first, a collection of adipose tissue; and,
second, by a peculiar formation of certain muscles. Dimples
generally form around joints, as well as on parts where the
soft, fatty tissue has accumulated. This latter class of dimples is found only on fat or plump persons. They indicate
ease-loving, mirthful, and affectionate natures. Dimpled
babes are always happy; having a great store of vital material, as shown by its producing dimples, gives ease and enjoyment to their existence. There are two sorts of dimples
which are produced by the peculiar formation of certain
muscles: The round dimple in the chin is a permanent feature, and does not depend upon the amount of adipose tissue
in the individual. It is caused by a peculiar formation of
the Levator Menti muscle. This muscle is located directly
on the front of the chin. This dimple is never seen where
the Bony system predominates, but always with the Muscular
system predominating over the Bony. This would bring it
into the formation in which artistic and literary talent is most
found. The meaning of this dimple is "love of the beautiful
of the opposite sex." It is frequently found in the chins of
talented persons in art and literature. Lord Byron, Dean
Swift, Richard Brinsley Sheridan, Mozart, Goethe, Jenny
Lind, Humboldt, Moliere, Horace Vernet, and a host of
other celebrities exhibit this peculiarity. The love of the
beautiful of the opposite sex is a part of creative talent; it
assists the sculptor and artist in forming their ideals, and if
this faculty did not reside in their own organisms on a large
scale, they could not reproduce in their creations the like
principle. Like not only begets like in offspring, but is able

to create similar effects in works of art. This dimple is often the sign of a voluptuous and pleasure-loving person, genererally good-natured and inclined to be generous. I name this the Louis XV. chin. That monarch was celebrated for his fondness for beautiful women; his chin shows this sign.

Dimples in the cheek are caused by a peculiar muscular formation, and the more they are called into action, of course, the more defined and permanent they become. Dimpled cheeks are often found where the Muscular system preponderates, and are always seen with large, full eyes in combination. This dimple is one sign for Approbativeness, and is deepened by much smiling. It is situated about one inch outward from the corners of the mouth. Several muscles come in close contact at this point, and this appearance is due to that circumstance and to the full development of all of them at this junction. This sign of Approbativeness is located next to Mirthfulness, and is its natural companion.

The dimples on the hands and about the mouth, shoulders, and other parts of the body, are caused by fatty tissue, and denote health, good nature, and, in fair-haired persons, amative disposition.

THE VOICE.

I think we may safely set it down as a fixed law of physiognomy that all those parts of the organism which depend upon the same organ-systems for their power exhibit unity of action and similar results in whatever part of the body their signs may be observed. Now, *straight* looks means straightness of action, whether this peculiarity is in the bones or in those organs which depend upon the interaction of the muscles to produce a straight, upright, and balanced condition. Eyes set straight in the head are more indicative of truth than eyes which are crooked; a straight mouth is more evidential of straightforwardness of speech than a one-sided one. The voice is produced mainly by action of the Muscular system; hence, every sign having relation to vocal expression is chiefly in that system. The eye, ear, larynx, mouth, tongue, nose, and lips are all involved in the use of the voice. The eye shows by its size the power for emotion, and therefore of the degree of expression which the individual possesses; the ear is a muscular organ; the vocal powers which reside in the throat are moved by muscular or cartilaginous fibres; the mouth also is surrounded by muscles, and is enabled to produced the varied expressions observed in talking, singing, shouting, and crying. From this unity of action we must infer a connection of purpose and method between these several organs and parts. Upon investigation this will be found to be the case. I have shown elsewhere that vocal expression is found best developed in those in whom the Muscular system predominates. In singing the individual must have a sufficient development of muscular power to produce all the wonderful vocal effects which we hear in such singers as Grisi, Malibran, Patti, Brignoli, and Tamberlik. In all the singers that have excelled this system will be found paramount. In those musicians who have taken rank as instrumentalists the same system is called into action; the muscles must be strong enough to *take command*, and be able to act independently of each other.

Many of our most magnetic orators possess this system greatly in the ascendancy. Beecher, Gough, and Ingersoll are types of this formation. I could swell the list, but these few examples suffice to call attention to this natural law of unity of action.

The *voice* is a great indicator of character and mental and physical power. The clearer the voice the clearer the mind; the sweeter the voice the more affectionate the possessor. Sensitiveness and sensibility are indicated by the voice. Some voices are "too sweet to be wholesome;" such are proofs of insincerity and deceit, or of secretiveness. Rough harsh voices denote strong harsh minds. Affected speech betrays the shallow or conceited individual; indeed, all affectations are assumed to cover deficiencies. One of the most belligerent women I have ever known had assumed a tone which was ludicrous in its softened affectation; the most casual observer would have known that the voice and face did not agree.

Rich, full, and rounded voices tell us that the muscular and adipose systems are well represented. Thin, nasal voices, with a sharp, disagreeable "twang," announce a lack of amiability and an impoverished condition generally, or else a deficiency in the Muscular and Vegetative systems. These give roundness and richness to the voice where they are rightly proportioned. With the Bony and Brain systems predominant the voice will be clear, strong, most decided, and energetic. Where the Vegetative system predominates there will be an undecided tone, with a good deal of "gab" and "chatter," with little sense, and no power except when in a passion, and then the voice becomes choked and shrill. Thus every form of the body is indicated by the voice. Sound is based on physiology and anatomy just as mind is, and can not be demonstrated without it. Study the voice by all means, and compare the individual with the voice. Very much of the character can be delineated by the voice, even by hearing it in the dark; that is to say, if the tones are natural, and have not been altered by disease of any kind.

Lisping voices betray a want of good balanced judgment; such tones are infantile in their nature, and, if found beyond the age of childhood, are to be classed in the category of enfeebled mentality in some direction. My experience of imperfections of the organs of speech (where they have originated by inheritance, and not by accident or disease) leads me to the conclusion that such defects disclose deficiency in truth-telling capacity or intellectual apprehension. Tongue-tied, lisping, stammering, or hesitating speakers certainly indicate enfeebled intellects or lack of moral powers. I speak now only of those who are by nature thus afflicted. As I have previously shown that any eye off the straight line evinces a crooked propensity, an untruthful, deceptive proclivity, so any naturally defective peculiarity of the mouth, tongue, or organs of speech discloses a similar condition of the moral or mental nature. The eye as well as the voice is controlled by the Muscular system mainly, and the same law governs every part of the same system. We cannot escape its deductions.

FINGER NAILS.

The finger and toe nails originated from the outer skin-covering, as the evolution of the organs of man teaches, and from this outer skin-covering were evolved the nerve and brain systems. The nails, like the skin and hair, must hence be indicative of the quality of the nervous system and mental activity. Upon investigation this will be found correct. The nails of those in whom the Brain system predominates will be smoother, finer, and thinner than those in whom other systems are more developed. This same principle holds good through the entire body. A thin skull indicates finer quality of mind than a thick one, and the term "numbskull" is quite apt when applied to dull persons. A rosy color of the nails implies a healthful condition, while pale or blue tints evince a weak constitution, an unbalanced circulation, or a functional disturbance. Nails which incline to bend over the tops of the fingers lead us to infer scrofulous and consumptive tendencies. Stiff and straight nails are found on firm, decided people. Soft, pliable nails belong to easy-going, yielding persons. Broad, square-cut nails are seen on the fingers of those who are more useful than ornamental; they denote more of the practical and mechanical than the artistic; hence, we must infer that they are to be found where the Bony system prevails; the possessors of such, being in the practical system, will be fond of the study of the truths of life, of science, of history and natural laws, and will be found inclined to reason and argument; will be clear-headed, and love debate and reform subjects generally.

The beautiful oval or almond-shaped nails are found on the hands of emotional people; those who love romance and ornamentation of all kinds; who have ardent feelings, refined tastes, rather inclined to be aristocratic in feeling, and, with large Credenciveness, are believers rather than investigators;

will enjoy ceremonial forms of religion, and be more inclined to speculative beliefs than those based on evidence and reason; will be of a naturally speculative mind if engaged in business. Many stock-dealers have this formation of the finger nails.

Rounding, small nails, rather thick and strong, always accompany round, thick, and strong bodies, full of vitality, with strong passions and emotions.

The law of harmony is illustrated in the nails as well as in all the other parts of the body. Nature never makes the mistake of putting artistic nails on a natural mechanic's hand, nor the nails of the Bony system on a round, plump, fat body. Nature is as harmonious here as in every department of man's complex organism. Her power is manifested as wonderfully in the adaptation of the nail to the finger as in the adaptation of the body to the brain. Indeed, I think the Creative Mind is more wonderfully illustrated in the world of the infinitely small than in the world of the infinitely great.

THE HANDS AND FEET.

The hands and feet must, according to the law of harmony, coincide with the shape of the body and brain; therefore, they are the exponents of mental, moral, and physical powers. The long bony hand, well supplied with muscles, gives us the clue to a mechanical and inventive mind. The hands and feet (where the muscles are predominant and supplied with some adipose tissue) that are short comparatively will reveal artistic and often literary abilities, with love of ease, pleasure, and social proclivities.

Very thin, colorless hands, almost transparent, denote a fragile, sensitive mind and body not long for this world, because the stock of inherited vitality is not sufficient to continue life to an extended period. When we see large hands that are so by nature, we infer the possessor to be powerful, especially if he has long arms and legs, with a broad mind, and free and generous with his resources, whether they be mental or material.

The short, thick, soft, fat hand belongs to the Vegetative system, and, of course, denotes the faculties which are found in that system. These are the domestic or social, with little firmness or reason, easy-going, inclined to stupidity and inertia, incapable of much mental effort, with no noble or lofty traits and aspirations, and much given to selfish enjoyment. A good observer can read character by the hand as well as by the face—not to such an extent, it is true, but enough to get the key to the entire organism. It is simply Comparative Anatomy put in practice.

The feet which correspond to certain shaped hands indicate similar traits, but the feet are so deformed by fashion that they are rarely seen in their natural conformation; and as the civilized races have the custom of covering the feet we cannot see them sufficiently to make much of a study of their peculiarities. Still, we know that they harmonize with the hands, body, and brain. A mechanical hand will have feet to match; that is to say, they will abound in bone and muscle. The artistic hand will accompany a muscular foot. The soft, thick, almost boneless, fat foot will tell us of an individual who loves ease, eating, drinking, and not very deep thinking; in short, a shining light in the "Fat Men's Club." Let every one of my readers make observations and comparisons for himself, and he will find the law of harmony in this direction exemplified in an astonishing manner.

WRINKLES.

Every appearance in the human face has a meaning and signifies something—stands for a sign. Now, wrinkles are in the highest degree evidences of character. Some of them express great beauty of character, others announce great and powerful intellect, while others still disclose character the most repulsive and wretched. As a general rule, deep wrinkles indicate a mind that has been immersed in profound thought and study. Surface people, those who "live on the outside of themselves," have very few wrinkles. Where I find a person past thirty years of age who has not formed some *creditable wrinkles*, I infer either a very shallow, selfish, unreasoning person, or a very deceptive, hypocritical one. There is an old saying that "gray hairs are honorable." We might say with more accuracy that wrinkles are honorable, provided they are in the right places and are the shape which denotes goodness or talent.

A smooth, shining, round face, without any wrinkles, belongs in an adult to a character suave, plausible, flattering, dishonest, and unprincipled, one who is "all things to all men." Such make good stock speculators and politicians, and are well calculated to get a living without working for it. Three or four deep wrinkles on the forehead which dip down in the centre are seen only on the foreheads of persons of good intellect. The face of Aristotle is a good example; his forehead as well as face shows very deep lines and wrinkles.

A number of confused or broken wrinkles on the forehead are signs of a confused understanding, a weak mind, and often of a weak body. Wrinkles which run obliquely outward from the corner of the eye are found on mirthful fun-loving people. Deep wrinkles under the chin and around the neck and wrists are evidences of too much fatty tissue, with the Brain system small. Such wrinkles belong to slow, easy-going persons, fond of the pleasures of the table, not inclined to intellectual labor, or any labor in fact. These wrinkles, when found on persons who have a large development of brain, give us characters of great mental vigor, and capable of hard and protracted mental labor. This would indicate the combination of the Mental and Vegetative systems. Dumas, the novelist, Hume and Gibbon, historians, Johnson, the philologist, Arkwright, the inventor, and other industrious brain workers, possessed this combination.

Numerous and very fine wrinkles all over the face, lying in every direction, show that the individual has passed a life of petty cares, of small savings, and has a fretful disposition. The same kind of wrinkles, where they are deep, reveal the miserly habits of a lifetime.

There are lines on the face which can scarcely be denominated wrinkles, but which are as indicative of character. A number of lines running down on the sides of the nose, which show very perceptibly while talking, are evidences of a petty, malicious, and dishonest character. Two or three wrinkles across the top of the nose (where the sign for Self-will is located) are indications that the individual has been used to the power of command habitually. Garibaldi's face exhibits this peculiarity. This sign is often seen on the noses of those who have superintended laborers and governed troops and sailors; sometimes on the noses of school teachers; also, parents who have good executive ability and will power. These wrinkles are caused by the exercise of these two faculties; hence, their location is where these two signs meet.

Nearly vertical wrinkles on the upper lip, running downward toward the mouth, exhibit patience, perseverance, and a spirit that has been obliged to submit without protest; also, ability to keep a "close mouth." The latter is a most useful quality to possess.

SMILES AND LAUGHTER.

An individual who is observed to have always a smile, simper, or smirk on his face, evidences an overweening degree of Approbativeness and desire to be approved of others, and this argues a want of independence of character—one who relies more on the opinion of the world than on his own conduct for satisfaction. Such characters are never great and broad, but show small capacity, and, by endeavoring to please every one, divert attention from their real character—or, rather, want of character—and so get judgment in their favor. So superficial is the estimate of the world, that foam, froth, and nonentity often excite more commendation than the most substantial qualities of character which do not present quite so attractive an exterior.

A loud, boisterous laugh belongs to a rude, unrefined person. A clear, mellow, ringing laugh, not too loud, announces
17

a clear-minded, harmonious character. The chuckling or suppressed laugh tells us that we have a secretive nature to deal with. Laughter which is spontaneous and full of merry tones, "like the jingling of sweet bells," discloses a frank, merry person, not yet spoiled by the world and the greed of mammon. A rude, short, loud "horse laugh" tells us of a most disgusting, rude, unfeeling brute. The hollow, affected laugh discloses an empty skull and a hollow heart. Such will do neither good nor harm to any one. A sharp, shrill laugh is evidential of a thin physique and an excitable temper, with an unbalanced and commonplace mind.

The laugh, like the tones of the voice used in speaking, is an unmistakable signification of sexual conditions and powers, as well as the exponent of other functional states. This fact assists still further in the proof before stated in this chapter; viz., that "all those parts of the organism which depend upon the same organ-system for their power exhibit unity of action and similar results." The proofs of the above-stated principle are, perhaps, better understood by the majority of people than many other physiological laws, because all persons have observed what is called the "change of voice," which is very marked in boys approaching the age of puberty. A change also takes place at this time in the voice of females, but is not so perceptible. This change of voice is correlated with a marked change and development in the reproductive system. Now, all the organs involved in reproduction are mainly muscular and fibroid, as are also the organs involved in the use of the voice.

Those persons who have been the most gifted in vocal expression, in song and oratory—such as our first-class opera singers, prima donnas, tenors, and bassos, and the great orators and elocutionists of the world—must have possessed sound and powerful reproductive systems. I believe that the record of their lives will bear me out in this statement, as well as the principle that creative art derives assistance from the procreative function. It is shown, in the "Evolution of Man," by Haeckel and others, that intelligence in the animal species did not progress greatly until a marked development of the reproductive system took place; and that, from that time on until man was evolved, the intelligence of the animal kingdom progressed accordingly as the reproductive system became gradually perfected. We cannot separate the mind from the body, nor mental faculties from physical functions; they are bound together by the God of Nature, and "what God has joined let no man put asunder."

There is another physiological fact known to the people generally, and that is, that as the sexual powers decline, the voice also loses its vigor and richness. From these observations, I think we are justified in considering the voice one indication of sexual conditions and powers, and the laugh, by its tones, enlightens us on this point, just as well as does the voice in speaking or singing. Still another proof of the connection between the remote parts of the muscular and fibroid system is had in the voices of eunuchs and in the soprano voices of the male singers in the Pope's choir. Emasculation in both cases has produced great and radical changes in the voices of these two classes of males. Any unprejudiced person can trace out these connections and correlations in the human organism—these which are so apparent to the senses. Most of the laws and principles laid down in this system of Physiognomy are so susceptible to demonstration by the senses alone, that one is hardly called upon, as Tyndall remarks, "to picture with the eye of the mind those operations which entirely elude the eye of the body." Observation and reflection, added to a love of truth and a candid mind, are all that are needed for this study.

UNLUCKY PERSONS.

There are in the world many persons who consider themselves, and who are considered by others, to be "unlucky;" that is, who are unfortunate in almost all their undertakings. My readers have doubtless asked themselves the question, Why is it that some people are always unlucky?—always in

trouble of one sort or another? There can be but one answer to this question: Such persons are lacking in some
department of their organization; there is either a mental
deficiency or a moral incapacity, or, as it often occurs, great
physical weakness of some part of the body, which always
culminates just in time to thwart all the well laid plans for
success. But whatever may be the failing, it may be taken
for granted that where "ill luck," as it is termed, follows one
through a life-time, the ill luck is caused by being ill constituted.

I have known, for example, a person whose nose was very
short—so short as to be entirely disproportioned to the rest of
the face. The character was equally inharmonious. Now, a
nose too short denotes, as you will find, a reckless, careless,
impudent, and imprudent nature. I am sure that to be all
this is to be "unlucky," indeed. This person, by careless,
reckless ways, lost her health, which, as she had not a balanced mind, she attributed to "ill luck." (This is a conveni-
ent phrase under which many seek to conceal their ignorance
and incapacity.) Then, in her intercourse with her friends,
she would converse with them in a manner which she thought
was charmingly "frank," but which they considered impudent, and often insulting. In this way she lost her friends
almost as fast as she gained them. Yet this person always
spoke of herself as "so unlucky," conveying the idea that
some unfortunate combination of circumstances was always
in waiting to injure her prospects, or that some power, outside of herself and beyond her control, was constantly assisting to make her "unlucky." Now you will ask, What was it
that caused this condition of her organism? As you have
read that all faculties are in their nature good and useful—
none bad or useless—why was it that this person should be
cursed, through no fault of her own, by an unbalanced organism, which always brought disaster to her and her friends?
This unbalanced condition, reader, was the product of a
combination of traits on the part of her parents, and which
should not have been perpetuated. Of course, they sinned

ignorantly in this matter; for, had they understood Scientific Physiognomy, they each might have married so as to have improved the race, and been the progenitors of more "lucky" children, because better endowed at birth with a balanced organism. Parents are censurable for creating evil conditions of mind and body, and transmitting them to their children.

Many persons are under the impression that marriage is holy under all circumstances, if only a minister or priest preside at the ceremony. This absurd opinion has arisen from the singular nomenclature used to designate certain laws and customs. A large class of persons, for example, believe that the laws and customs of sectarian and theological societies are divine, and that divinity belongs exclusively to this class of laws and beliefs. Now, this is an absurdity; for every law which relates to our physical nature is just as divine as the laws which govern sectarian and theological societies, and really of far more importance, as broken physical laws lead to immorality and the grossest irreligion. When we can comprehend physical religion, and strive to live up to it, "unlucky" persons will cease to be perpetuated, and true religion will give strong, well balanced, moral, intellectual, and happy organisms, with moral power and brain power, added to such health as will enable them to command circumstances. Such persons will never prove unlucky to themselves or their friends. There are many just such unlucky persons in our penitentiaries as the one instanced. An examination of the physiognomies of these unfortunates will prove that many of them have inherited unbalanced, dwarfed, perverted, or one-sided natures. The physical and moral sins of the fathers and mothers have been visited upon their children; and although, for the safety of the community, these "unlucky," unfortunate individuals must be pent up, yet we should, in justice—not charity, but in a spirit of justice—overlook much of their evil-doing, and censure them only for not trying to eradicate the evils implanted in their natures by a law which is more just and divine in its operations than the ceremony which gave their progenitors license to perpetuate such moral monstrosities. "I will repay, saith the Lord." Broken physical laws are just as great an abomination in the sight of God as broken moral ones are; they are inseparably bound together, and cannot be judged apart. And whether parents sin ignorantly or sin willfully, Nature takes no notice of the motive, but inflicts the penalty with the most exact justice, because natural or divine law, whichever you may term it, is unerring, and no respecter of persons.

CHAPTER X.

ORIGIN AND EVOLUTION OF THE ORGANS.

"Systematic Physiology is based especially upon the history of development, and unless this is more complete, can never make rapid progress; for the history of development furnishes the philosopher with the materials necessary for the secure construction of a system of organic life. We should study each organ, each tissue, and even each function, simply with the view of determining whence they have arisen."—HUSCHKE.

In viewing the beautiful ideals of art, as shown by the sculptured marbles of the great masters of ancient Greece; in regarding the beauteous blending of color and imagery, as exhibited in the works of Titian, Correggio, and Michael Angelo; in beholding the grand and sublime efforts of some impassioned orator; or in contemplating the wonderful mechanism brought into existence by the creative mind of a master inventor—the thoughtful observer will, no doubt, ask himself these questions: How has man become possessed of the varied powers necessary to the perfecting of all these creations?—did he come come into existence fully equipped, like Minerva from the brain of Jupiter, and endowed with all the faculties essential to these operations? No, reader !

NATIVE AUSTRALIANS.
Showing a lack of Muscular Development.   (After M. d'Urville.)

Nature produces no miracles; in her domain, orderly, progressive, unerring, infallible law is the method by which perfection is attained.

This is as true in the department of Sociology as of Biology, and governments go forward only as fast as the people are prepared. This preparation is also a matter of growth and development, and society moves forward on fixed lines, presided over by immutable laws. There are no miracles in Nature, and no retrogression. All who have gazed upon the pictured representations of the native Australians, or those who have visited them in their own habitat, have, no doubt, observed in their organisms the absence of certain faculties and certain functions. The muscular system is seen to be very defective, as evidenced by the lack of muscular development in the calves of the legs, in the arms, and in the entire body. The faculties which derive their power from this system are, consequently, lacking in this people. The

architectural, artistic, literary, and mathematical powers are
entirely wanting. Their rude habitations cannot compare, in
architectural skill, with the buildings of the ant, wasp, mole,
or beaver. You may say that they have speech, and that
speech is a "divine gift." How, then, does it occur, if these
people are "divine"—the children of God, above and beyond
all the lower animal creations, and endowed with the so-called
"divine gift" of speech—I ask, how does it occur that these
people are so undeveloped as not to compare, in natural intelligence, with some races of dogs, for example?

Is it because they have had no education—no schools,
churches, hospitals, and jails, and other concomitants of
civilization? Not at all; these would be as useless to them
as clothing, houses, and furniture. They are incapable of
further development. Their language alone would prevent
their progress, since it is, like themselves, in its infancy;
they speak in guttural monosyllables, like babes when they
first essay speech. Theirs is a condition of arrested development; they have become paralyzed—ossified; they can go
no further, and will die out; for when progression ceases,
annihilation results. The conditions requisite for the growth
of this race were absent. For ages they lived without admixture of other blood, and this alone produces stagnation.
They occupied what may properly be termed an island home,
where no great beasts ranged to invite the force of man to
their destruction. This one circumstance prevented their
advance by impeding muscular development; and without
the development of the muscular system, the grandest
achievements of civilization are impossible.

Those who have followed the course of this system of
Physiognomy will have seen how many beautiful faculties
are evolved from the Muscular system—how many depend
upon its perfection and dominance. Mechanism, art, commerce, sentiment, and social life find in the high development
of the Muscular system their best illustration. Blot out
from the human organism all these, and what remains? An
organism incapable of further evolution; because Nature

never leaps, and cannot progress except in her regular order.
If the muscles have not been properly developed, the functions and faculties which are related to the Muscular system
will not make their appearance, and the Bone and Brain
systems will not be perfected. Hence, destruction will follow
any race that does not move according to the laws and requirements of evolution.

How do we know the methods of Nature in regard to
man's evolution? There are three sources from which we
derive this knowledge: Comparative Anatomy, Physiology,
and Embryology. These sciences write the history of man's
development in living letters. The means of obtaining this
knowledge are in existence in the world at the present time.
The profoundest minds of the age are turned to the investigation of the origin of man and of his mind; it is reasonable
to conclude that they will bring forth results of their investigations in the shape of proofs. This they have already
done to some extent; and, by the vast researches of one man
alone—Ernst Haeckel—we are able to trace the evolution of
races, and of the organs of animal and human organisms.

My theories of the nature, origin, location, and meaning
of the several organs, functions, and faculties of the human
mind and body have been shaping for years. I have re-
frained from putting them forward, because I knew that, on
account of their novelty, they would be subjected to severe
and adverse criticism; because, also, I had not the *corroborative testimony* of those better known to science. But, as
time has progressed, investigation and research on the part
of eminent thinkers have given me all the evidence I need to
sustain the basilar principles of my system. Observation of
the faces and forms of men, women, and animals will supply
the rest.

The more I investigate Mr. Haeckel's system of evolution,
the more profoundly am I impressed with its truth. In it I
find the corroborations of my system, or at least many parts
of it; and I blush while I write it, that one so obscure as
myself can claim to be able to corroborate anything that so

great a mind has advanced. I also find in Physiognomy the corroboration of much that he has stated, but especially have I found in the human face the proofs of the evolution of the organs and systems of the body of man, simply by the order of the location of their signs in the face. The order of their placing and action in the body is also a proof of many of his positions, and mine as well.

Here I am about to attempt a very difficult task. I am desirous of giving my reader somewhat of Mr. Haeckel's system of the origin and progress of the systems and organs of the body in a few pages. This is a work to which he devotes two volumes, and my attempt may be thought presumptuous; but still it is my proof, and I hope I may accomplish my task without injustice to his elaborate descriptions, illustrated, as they are, with numerous diagrams and plates. It must be borne in mind that the knowledge of the origin and progressive development of the entire man, as he now stands perfected, has been sought for, first, in the simplest organisms in the world—viz., the amœba; thence coming along up the scale of progressive evolution to the family of worms; thence along the line of investigation to the brainless fishes; thence to the skulled organisms, through reptiles, birds, and beasts, to man.

In order to make my evidence more complete, I will go back to primeval times and take up the investigation of primeval organisms; because, in their origin and evolution, they type the growth and progress of man, his organs and functions. The amœba is composed of a small speck of slime, or plasmoid substance. Under the microscope it discloses a simple cell, or germ. This form is the beginning of life in every plant, animal, and man in the world. Man, at his commencement, is nothing more than this—a small cell, or germ, combined with a microscopic quantity of mucus-like substance. In the case of the amœba, we find that, without any organs, it yet has powers such as are seen only in developed organisms. It seems to possess the faculties of motion without muscles, bones, or limbs, irritability without nerves, digestion without stomach, reproduction without sexual organs, and respiration without lungs; and, withal, purely chemical in its action. At the first dawn of all things in existence, chemical action alone seems to be dominant. As it lies in the water—for this is its natural abode, and it can be seen in bodies of both fresh and salt water almost any day by seeking for it—it can project one part like a limb; it can expand, contract, or roll up in a spherical shape. It digests by absorption the minute animalculæ contained in the water, thus showing it to be carnivorous without teeth. It reproduces by fission, or division; that is to say, after it has attained a certain size it separates into two parts, and these again, in their turn, repeat the process when the right proportions are reached.

You will ask how it is possible for an animal to do all this without organs or functions. The answer I make to this, is that these powers must be *diffused* through the creature just as they are at the other extreme of evolution; just as they are found in man, the highest expression of organized life, as the amœba is the lowest; just as all the elements of life are *diffused* through oxygen, hydrogen, nitrogen, and carbon. All the possibilities of organic form and life are in these simple constituents; yet we see that they have neither form nor organs. Although man has several systems of functions, still they are so blended and interrelated that it is impossible for any one of them to act independently; they are *diffused*, so to speak, all through the human system. The nerves, the muscles, the bones, the tissues, the mind itself even, is diffused, and irritation and sensibility proceed from all the nerve-ganglia in the body, just as irritation and sensibility are manifested by the microscopic amœba, without any perceptible nervous system.

The next stage of progression after the amœba is a simple aggregation of cells, without organs as yet. The manner in which these germ-cells aggregate or break up into other cells is most interesting, and has been observed in very low mammals, in guinea pigs and rabbits, in the amphioxus, and in the eggs of toads and frogs. The eggs of all these animals develop exactly as do the cells of the amœba. Observation of the manner in which the eggs of frogs develop shows that their eggs are circular, and the upper half appears darker than the lower; this marks the egg into two distinct halves. This marking into halves commences about an hour after being deposited; an hour later another line or furrow is formed, cutting the first at right angles. This change continues in geometrical progression, from two to four, thence to eight, to twelve, to sixteen, to twenty-four, to thirty-six, to forty-eight, to sixty-four, until one hundred and sixty cells are formed, the greater number of which consist of the cells which later form the animal functions; the less number, the vegetative functions of the animal. This law of mathematical progression is one proof of my proposition that all the operations of Nature have mathematical law as a common basis. The commencement of all life is on so infinitesimal a scale that, until the microscope reached its present perfection, the means to ascertain the laws of evolution did not exist.

The next stage appears as a simple hollow globe filled with liquid, the wall of which consists of a single layer of cells.

The next progressive step shows us a hollow body with an opening at one end, the wall consisting of two different cell-strata. These strata Mr. Haeckel describes thus: "The two cellular layers which surround the cavity of the primitive intestine, and alone constitute the wall of the latter, are of very great significance; for these two, which alone constitute the whole body, are, in fact, the two primary germ-layers, or primitive germ-layers. The outer cell-layer is the skin-layer, or exoderm; the inner cell-layer is the intestinal layer, or entoderm. The whole body of all true animals proceeds solely from these two primary germ-layers. The skin-layer furnishes the outer body-wall; the intestinal layer forms the inner wall of the intestine, and directly surrounds the intestinal cavity. At a later period, a cavity forms between the two germ-layers; this cavity, filled with blood or lymph, is the body cavity (cœloma). The two primary germ-layers—the outer, or serous, and the inner, or mucus layer—were first clearly distinguished in 1817, by Pander, in the incubated chick; but their full significance was first thoroughly recognized by Baer, in 1828, who gave the name of animal layer to the outer layer; that of vegetative layer to the inner. These names are very apt, because it is the outer layer which especially, if not exclusively, gives rise to the animal organs of sensation and movement—the skin, the nerves, and the muscles; while, on the other hand, it is especially from the inner layer that the *vegetative* organs of nourishment and reproduction—the intestine and blood-vessel systems—arise."

This hollow body mentioned above is the primitive intestine; the opening, the first appearance of the mouth; the two different kinds of cell-strata form the inner and outer skin; the inner skin assists in digestion, and the outer forms the covering and assists in motion and sensation.

The next advance made shows an organism—the turbellaria, a gliding worm, which is found at the present day in both fresh and salt water. These creatures have two openings to the body, a nerve system consisting of a simple nerve ganglion at the top of the mouth-opening, a pair of simple eyes, and nose-pits; also, will be found a pair of simple kidney ducts. Mr. Haeckel remarks:

"The appearance of these (kidney ducts) at so early a period shows that the kidneys are very important primordial organs. It also shows their existence in all flat worms; for even the tape-worms, which in consequence of the adoption of a parasitic mode of life have lost the *intestine*, yet have the two secreting *primitive kidneys*, or excretory ducts. The latter, therefore, seem to be older and of greater physiological importance than the blood-vessel system, which is wholly wanting in the flat worms."

The reader will observe that the kidney system makes its appearance before the heart, liver, lungs, blood circulating system, brains, bones, or any of the smaller organs or systems of the body. Mr. Haeckel observed this fact, and, as he expresses it, "the kidneys seem to be of greater physiological importance than the blood-vessel system." Of the first appearance of the kidney system, Mr. Haeckel observes:

"These four organ-systems which have been mentioned were already in existence when an apparatus developed tertiarily in the human ancestral line, which at first sight seems of subordinate significance, but which proves, by its early appearance in the animal series and in the embryo, that it must be very ancient, and consequently of great *physiological* and *morphological* value. This is the urinary apparatus, or kidney system, the organ-system which secretes and removes the useless fluids from the body."

We have already seen how the primitive kidneys appear in the embryo of all vertebrates long before any trace of the heart is discoverable. Later on Mr. Haeckel remarks:

"The human skin and intestines are, according to this, many thousands of years older than the muscles and nerves. These again are much more ancient than the kidneys and blood-vessels, and the latter, finally, are many thousands of years older than the skeleton and the sexual organs. The common view that the vascular system—that is, the blood circulating system—is one of the most important and original organ-systems is therefore erroneous. It is as false as the assumption of Aristotle that the heart is the first part to form in the incubated chick. On the contrary, all the lower intestinal animals show plainly that the historic evolution of the vascular system did not begin till a comparatively late period."

My observations in my own peculiar branch of science lead me to see the importance of the kidney system, not only from a physiological standpoint, but also from a moral one. The chapter on the "Rationale of Physical Functions and their Signs in the Face" explains this theory.

The two systems of organs which appeared first in man's primitive ancestors were the intestinal and skin systems; these came simultaneously. After these came the gill-intestine, which foreshadowed the lungs. A rudimentary stomach was also evolved. The two systems which appeared next in order and simultaneously were the nerve and muscle systems. Then evolved the blood circulating system, as yet without a heart, the blood circulating through tubes without any central organ.

The next set of systems which appeared were, first, the skeleton, and then the sexual system; reproduction previous to this having been produced by fission, or in other ways not requiring sexual organs. The first two systems which appeared were, as above stated, the intestinal and the outer skin-covering, which were used for motion, and also for sensation. This sense of touch stood in place of nerves to these low organisms; they gained all knowledge of their surroundings from the sense of touch, and "without touch," says Taine, "nothing could exist."

Later on in evolution, this outer skin, which "had become *especially sensitive*, gradually withdrew into the shelter of the interior of the body, and there laid the first foundation of a central nervous organ. As differentiation advanced, the distance and distinction between the external skin-covering and the central nervous system detached from this became continually greater, and finally the two were permanently connected by the conductive peripheric nerves."

"Let us now," says Haeckel, "turn aside from these very interesting features in evolution, and examine the development of the later human skin-covering, with its hairs, sweat glands, etc. The skin, in the first place, forms the general protective covering which covers the whole surface of the body, and protects all other parts. As such, it at the same time effects a certain change of matter between the body and the surrounding atmospheric air— perspiration, or skin-breathing. In the second place, the skin is the oldest and most primitive sense-organ, the organ of touch, which effects the sensation of the surrounding temperature, and of the pressure or resistance of bodies with which it comes in contact. Those organs of our body which discharge the highest and most perfect functions of animal life—those of sensation, volition, thought; in a word, the organs of the psyche, of mental life—arise from the external skin-covering."

The corroboration of this last sentence of Mr. Haeckel is found stated in one of the sub-basic principles of scientific physiognomy—"Texture is significant of quality"—for, without seeing the face of an individual, or his form even, the *quality* of his mentality is disclosed by the quality of his skin and hair, both of which must and do correspond always to man's mental quality. The finer, clearer, and more sensitive the skin, the finer will be the quality of the mental sensations and sensibility, or, in other words, of his brain and nerves. Thus another proof of my propositions is given us from this great man's research. Although the two sciences on their first presentation do not seem to be directly connected, yet as we proceed we shall find that they are corroborative of each other.

In Physiognomy the Brain and Nerve system is located the highest in the organism, and comes last in the order of progressive development, for the reason that the true brain, the perfected cerebrum, was the last organ in developing, and is the chief seat of mentality, although mentality is diffused through all of the several systems of the entire organism, whether of man or animal.

The first appearance of anything like a skeleton is the notochord, which is not yet true bone, but cartilaginous in its nature. It foreshadows the vertebræ, or what is commonly called the backbone. Along the inner side of this cord a medullary or nerve-tube is found, which has evolved from the upper throat ganglia—the first appearance of a nervous system. This notochord develops sufficient strength later on in evolution to support strong side-muscles and an oar-like tail, which were needed for swimming. From the anterior portion of the notochord, near and above the mouth-opening, a little capsule made its appearance. This is the first beginning of a brain. Hitherto the mental powers of animal organisms, their consciousness and sensibilities, the sense of feeling or touch, have been located in the body, in the inner and outer skin, in the muscles and notochord.

Now a great step forward is taken. A single nostril forms above this capsule, and nostrils presuppose a use. At the side of the animal in which this stage of evolution exists, just below a simple eye which has formed, are seven little openings, called gill-openings. The air contained in the water taken in at the mouth, which is only a round opening without jaws as yet, is respired through these little gill-openings; this is the first approach toward breathing through an apparatus especially for that purpose. Hitherto breathing has been carried on by lower organisms through a process called skin respiration, or by using the oxygen contained in the water taken into the mouth-opening.

These characteristics are all of the most important which have evolved from the first simple one-celled amœba until the fish family is reached. Heretofore we have considered very low organisms—gastræa, worms, and lampreys. The evolution of fishes marks a great advance in the origin of organs. The one little capsular brain has formed four other similar little bladders, which, later on, form one whole brain; these five parts are the origin of the five parts of the brain as they are found divided in the human skull. Two jaws also appear, two nostrils, and the swimming bladder, which organ develops into the true lung in the higher vertebrates; as now found in the fish it is used as a hydrostatic apparatus, by means of which the fish rises and sinks in the water. The swimming bladder is developed from the anterior portion of the intestinal canal, and corresponds in its position to the lungs in the higher organisms.

The strong side-muscles which were evolved in the swimming worm now develop into two fore and two hind limbs; the two fore limbs are called pectoral fins, and the hind limbs ventral fins. These fins foreshadow the upper and lower limbs, the hands and feet, of man, and the limbs of all vertebrate animals. With the coming of these there appeared a sympathetic nerve system, a spleen, and salivary gland. In this stage of progress the notochord has ossified and become true bone, although fish-bone is always more like cartilage

16

than the bones of higher animals. Some little bony arches, called gill-arches have been thrown upward and forward from the anterior portion of the notochord, or, as it now is, the backbone of the fish, and these form the upper and lower jaws. Fishes have from four to six pairs of gill-openings, which lie between the gill-arches. "In the embryos of man and the higher vertebrates only three or four pairs are developed." In the latter only a single vestige of a gill-opening remains, the remnant of the first gill-opening. This changes into a part of the organ of hearing; from it originates the outer ear-canal, the tympanic cavity, and the Eustachian tube.

In all the three higher vertebrate classes, also in man, the tongue-bone (os hyoides) and the bonelets of the ear originate from the gill-arches. From the first gill-arch, from the centre of the inner surface of which the muscular tongue grows, proceeds the rudimentary jaw skeleton, the upper and lower jaws, which inclose the cavity of the mouth and carry the teeth. The original formation of the human mouth-skeleton of the upper and lower jaws can thus be traced back to the earliest fishes, from which we have inherited them.

The next stage of evolution brings us to the amphibia—creatures endowed with the power to live on land or in water. In this class of animals a lung for breathing while upon land is required. This organ evolves from the forward and upper end of the intestine, and the air is inhaled through a tube, or windpipe. "At the upper end of the windpipe, below its entrance into the throat, the larynx, the organ of voice and speech, develops. The larynx occurs even in amphibia in various stages of development, and with the aid of Comparative Anatomy we can trace the progressive development of this important organ from its very simple rudiment in the lower amphibia up to the complex vocal apparatus represented by the larynx of birds and mammals."

The power for breathing necessitates a heart, or a blood circulating system. We accordingly find in the order of the amphibia a heart, not yet perfected in its power and mechanism as in the higher organism of reptiles, birds, and mammals. Of this change in the mode of breathing air directly from the atmosphere instead of from water, Mr. Haeckel says: "This physiologically significant modification of the mode of respiration is the most influential change that affected the animal organism in the transition from water to dry land. In the first place, it caused the development of an air-breathing organ, the lung, the water-breathing gills having previously acted as respiratory organs. Simultaneously, however, it effected a remarkable change in the circulation of the blood and in the organs connected with this, for these are always most closely correlated with the respiratory organs."

In this last sentence the reader will find the proof of the origin of my sign for the action of the heart: "The larger the lung and nostril the greater the size and power of the heart." These two organ-systems are closely interrelated, and one always conditions the other. The change from water-breathing to air-breathing led to many other important changes. Of this transformation Mr. Haeckel remarks:

"Within the vertebrate tribe it was undoubtedly a branch of the primitive fishes (*Selachii*), which, during the Devonian period, made the most successful effort to accustom itself to terrestrial life, and breathe atmospheric air. In this the swimming bladder was especially of service, for it succeeded in adapting itself to respiration of air, and so became a true lung. The immediate consequence of this was the modification of the heart and *nose*."

Here is still another proof of the origin of my sign in the the face for the power and activity of the circulatory system and heart. Let us return to the further description of the evolution of the organs by Mr. Haeckel. He says:

"While the true fishes have only the blind nose-pits on the surface of the head, these now become connected with the mouth-cavity by an open passage; a canal formed in each side leading directly from the nose-pit into the mouth-cavity, and thus, even while the mouth-opening was closed, atmospheric air could be introduced into the lungs. While, moreover, in all true fishes the heart simply consists of two compartments—an auricle which receives the venous blood from the veins of the body, and a ventricle which forces this blood through an arterial expansion into the gills—the auricle, owing to the formation of an incomplete partition-wall, is now divided into a right and left half; the right auricle alone now received the venous blood of the body, while the left auricle received the pulmonic venous blood passing from the lungs and the gills to the heart. The simple blood circulation of the true fishes thus became the so-called double circulation of the higher vertebrates, and this development resulted, in accordance with the laws of correlation, in further progress in the structure of other organs.

"The vertebrate class which thus first adapted itself to the habit of breathing air is called mud-fishes—*dipneusta*, or double-breathers, because like the lowest amphibia they retain the earlier mode of breathing through the gills in addition to the newly acquired lung respiration. In their mode of life they are true amphibia. During the tropical winter, in the rainy season, they swim in the water like fishes, and inhale water through the gills. During the dry season they burrow in the mud as it dries up, and during that period breathe air through lungs, like amphibians and higher vertebrates."

The life on land of these amphibious creatures necessitated on apparatus for locomotion. This caused an advance in the strength of the side-muscles which were attached to the fins, and a change in the fins themselves. Of the construction of these fins, Mr. Haeckel observes:

"The thorough researches of Gegenbaur have shown that the fins of fishes, concerning which very erroneous views were previously held, are feet with numerous digits; that is to say, the cartilaginous or osseous rays, many of which occur in every fish-fin, correspond to the fingers or digits on the limbs of higher vertebrates; the several joints of each ray correspond to the several joints of each digit. In the double-breathers the fin yet retains the same structure as in fishes, and it was only gradually that the five-toed form of foot, which occurs for the first time in amphibians, was developed from this multidigitate form.

"The great significance of the five digits depends on the fact that this number has been transmitted from the amphibia to all higher vertebrates. It would be impossible to discover any reason why in the lowest amphibia, as well as in reptiles and in higher vertebrates up to man, there should always originally be five digits on each of the anterior and posterior limbs, if we denied that heredity from a common five-fingered parent-form is the efficient cause of this phenomenon. Heredity alone can account for it. In many amphibia certainly, as well as in many higher vertebrates, we find less than five digits, but in all these cases it can be shown that separate digits have retrograded and have finally been completely lost. The causes which led to the development of the five-fingered foot of the higher vertebrates in this amphibian parent-form from the many fingered foot, must certainly be found in the adaptation to the totally altered functions which the limbs had to discharge during the transition from an exclusively aquatic life to the one which was partially terrestrial. While the many-fingered fins of the fish had previously served almost exclusively to propel the body through the water, they had now also to support the animal while creeping upon land. This effected a modification both of the skeleton and of the muscles of the limbs. The number of fin-rays was gradually lessened, and was finally reduced to five. These five remaining rays now, however, developed more vigorously. The soft cartilaginous rays became hard bones; the rest of the skeleton became consequently more firm; the movements of the body became not only more vigorous but more varied. The separate portions of the skeleton system, and consequently of the muscular system, also became more and more differentiated, owing to the intimate correlation of the muscular to the nervous system; the latter also naturally made marked progress in point of functions and structure. We find, therefore, that the brain is much more developed in the higher amphibia than in mud-fishes and in the lower amphibia."

In the last few sentences of this description of the evolution of the amphibia I find the corroboration of my theory of the progressive development of some of man's physical functions and mental faculties. I have shown elsewhere that as the muscular system evolved and became differentiated man's capacity for mental progression was enhanced. If the reader will refer to the opening pages of this chapter, he will find in the illustration I make of the native Australians how exactly in accord my theory of the order of development of the functions in man is with Mr. Haeckel's proofs of the evolution of the organs and functions, not only in the lowest organism, but in the embryonic life of man, and lastly in his most developed and perfected state as a full grown member of the latest and highest race.

I cannot give within the limits of this work (neither is it pertinent) the history of the evolution of all the various organ-systems in the human body, I shall present simply those which in my judgment seem to be corroborative of theories which are the result of years of observation and research on my part. Without pretending to carry the reader along the regular course of the evolution of the organs, so as to show the development of the structure of the skull, brain, and interior organs of the body not already noticed, I shall content myself with giving some extracts from different parts of the second volume of Haeckel's "Evolution of Man," bearing upon and sustaining my theory of the origin and location of mind. I will first quote a paragraph which is simply a reiteration of what has already been adduced by Flourens, Longet, and other well known anatomists. Mr. Haeckel remarks:

"It is possible to remove the great hemispheres of a mammal piece by piece without killing the animal, thus proving that the higher mental activities, consciousness and thought, conscious volition and sensation, may be destroyed one by one, and finally entirely annihilated. If the animal thus treated is artificially fed, it may be kept alive for a long time, for the nourishment, digestion, respiration, the circulation of the blood, the secretion—in short, the *vegetative functions*—are in no way destroyed by the destruction of this most important mental organ. Conscious sensation and voluntary motion, the capacity for thought, and the combination of the various higher mental activities, have alone been lost."

Of the origin of the source of mind, sensation, and consciousness, he says:

"Comparative Anatomy and Physiology show that in the low animals specialized sense-organs are entirely wanting, and that all sensations are transmitted through the outer surface of the skin-covering. The undifferentiated skin-layer, or exoderm, of the gastræa is the simple cell-layer, from which the differentiated sense-organs of all intestinal animals (*Metazoa*), and therefore of all vertebrates, originally developed. Starting from the consideration that necessarily only the most superficial parts of the body, those immediately exposed to the outer world, could have accomplished sensations, we should be justified in conjecturing *a priori* that the organs of sense also owe their origin to the same source. This is indeed the fact."

Elsewhere Mr. Haeckel observes: "The history of evolution, in conjunction with the rapidly advancing comparative anatomy and physiology of the sense-organs, affords the only safe foundation for the natural theory of the mind."

Speaking of the varying degrees of intelligence or mental activity in the lowest vertebrates, he remarks:

"Side by side within the various classes, orders, genera, and species, we find so great a variety of vertebral intellects that at first sight one can scarcely deem it possible that they can all be derived from the mind of a common primitive vertebrate. First, there is the little lancelet, which has no brain at all, but only a simple medullary tube, the entire mental capacity remaining at the very lowest grade occurring among vertebrates. The cyclostoma also, standing just above, exhibit a hardly higher life, though they have a brain. Passing on to the fishes, we find these intelligences, as is well known, at a very low point. Not until from these we ascend to the amphibia is any essential progress in mental development observable."

Here, again, I find still further proofs of my theories. The reader will remember that I have stated elsewhere that the natural order of the progressive development in the races of man, from the lowest to the higher, was from the Vegetative system to the Thoracic, from the Thoracic to the Muscular. Now, the great advance made by the amphibia was in the increased development of the Muscular system; first, in the addition of lungs, heart, and muscles, to assist locomotion on land. This increased action of the Muscular system, of course, advanced the mental powers and activities. The amphibia would necessarily be brought into contact with new and diverse methods of life, in order to establish itself on land, and to accommodate itself to the new conditions which this form of life entailed. The circumstance of having to provide food other than that supplied by the waters would not only strengthen and cultivate the Muscular system, but would sharpen the mental activities of the creature. Its life being passed partly on land would strengthen the Bony system, for sunlight is essential to the development of the osseous structure in animals as well as man. And here the next great advance in physiological and anatomical and mental development was made; for the reader will please observe that faculties and functions advance and develop simultaneously.

Let us here continue Mr. Haeckel's description of the evolution of mind in the lower vertebrates. He says: "This progress in mental development is much greater in mammals; although, even here in the beaked animals *(Ornithostoma)*, and the next higher class, the stupid pouched animals *(Marsupials)*, the entire mental activity is still of a very low order; but if we pass on from these to *placental* animals, within this multiform group we find such numerous and important steps in differentiation and improvement, that the mental differences between the most stupid placental animals (for instance, sloths and armadilloes) and the most intelligent animals of the same group (for instance, dogs and apes) seem much more considerable than the differences in the intellectual life of dogs, apes, and men."

This last mentioned advance, my readers will observe, is based on the differentiations which resulted in a more complex arrangement of the organs and functions of reproduction, and of all the concomitant functions and faculties which this great advance requires. A more extensive and complex nervous system and brain would be necessitated by such change, and in the placental animals we accordingly find that the mentality has advanced in the ratio of their physical and anatomical development. Herein I find another proof of one of my theories; namely, that functions and faculties are correlated; that the mental cannot progress without the physical powers; that they depend upon each other, condition each other; that, in short, mind is a part of the *entire body*, and does not inhabit any one particular portion of the organism, but is diffused all through it—is blended with every function, and is part of every function. This knowledge simplifies the doctrine of mind, spirit, and soul very materially. As the mind or brain has always been considered the organ of the spirit or soul by theologians and metaphysicians, Comparative Anatomy will give them all the evidence needed to ascertain its locality and attributes. Of the difference between nature and spirit, Mr. Haeckel observes:

"Accordingly, we cannot assent to the popular distinction between nature and spirit. Spirit exists everywhere in nature, and we know of no spirit outside of nature; hence, also, the usual distinction between natural science and mental science is entirely untenable; every real science is at the same time both a natural and a mental science; man is not above nature, but in nature."

In closing this review of portions of Mr. Haeckel's "Evolution of Man," let us compare the points of resemblance and correspondence between the two sciences, and summarize the proofs by which his evidence is corroborative of my discoveries in scientific physiognomy. The first point of

resemblance between the two is evidential of one of my basilar laws; namely, that all creations, as well as all particles
of matter, have for their basis three underlying laws—those
of chemistry, architecture, and mathematics. I have shown,
in the description of germ-cells, that all life is at first a
simple cell, a purely chemical compound; in its next stage
it takes on a fixed and definite form or shape, thus showing
its architectural proportions. The number of the divisions
of the germ-cells in geometrical ratio proves that mathematical law governs every particle of matter, and controls the
physical as well as mental basis of life.

My theory of the high importance of the kidney system as
a moral agent and a purifier of the body and mind, I think,
is well sustained by the investigations of Mr. Haeckel among
the lower animal organisms, where he finds the existence of
kidney ducts long before any of the organs and functions,
which are considered by the generality of people more essential to the existence of the human organism. Of course,
physicians and physiologists comprehend thoroughly the
"high physiological importance," as Mr. Haeckel terms it,
of the kidney system; because they know that, whereas the
functions of the bowel system can be suspended for from
twenty to thirty days without causing death, the suspension
of the functions of the kidney system will cause convulsions
and death in almost as many hours.

Well might it be said that the early appearance of this
system in the lower organisms showed it to be "of great
physiological importance." It is not equaled by any other
system in the body in point of necessity and importance.
The general belief is that the bowel system is the greatest
excretory power of the body. This is not correct; the skin,
which is closely related to the kidney system, far exceeds the
bowel system in importance as an excretory agent. This
intimate relation of the skin and kidneys is proved, by the
investigations of Mr. Haeckel, by the fact of the kidneys
having evolved from the outer skin sensory layer. The
every-day experience of almost all adults proves this relation-

ship between the skin and kidneys; for where the pores of
the skin have become inactive by reason of chill or cold, the
kidneys act for them, and throw off an increased amount of
waste material, and this action is carried on *vice versa*.

That the nerves, brain, and kidneys have originated from
the same source—that is, by evolution—is proved, not only
by Mr. Haeckel's observation, but by the facts and experi-
ences of life; by the joint indications and signs of all these
functions in the bodies of human beings. The finer and
clearer the skin, the finer the grade of mentality is found to
be.   Compare, for example, the texture of "Sitting Bull's"
skin with the finely organized cuticle of Elizabeth Barrett
Browning, the grand poetess. Another proof of their common
origin is found in the fact that a finely organized skin assists
in carrying off the waste and impurities of the body, thus
assisting the kidneys in excreting impurities which lead to
immorality where they are not discharged; while, at the same
time, this finely organized skin is an indication of purity of
thought, which characterizes all who have a fine quality of
brain, or where the Brain and Nerve systems predominate
over all other systems of the organism; thus proving my
theory that Conscientiousness is related to the kidney system.
I do not think that this position can be controverted, except
on the old theological basis that mind is something apart
from the body, and governed by a "soul," the location and
qualities of which have never, to my knowledge, been ascertained.

The next point of resemblance, and corroboration of my
sign for the size of the nostrils and lungs, and the corresponding strength and vigor of the blood circulating system, are found in the simultaneous appearance, in the low
fish organisms, of these three organs and systems. Every
indication of the human face and body proves the correctness
of these signs. Wherever the nostrils are wide and large,
or round and large, the lungs correspond in size and shape.
The heart, also, must necessarily be of large size and of
powerful action, in order to receive all the blood which large

lungs oxygenate. Thus these systems are naturally and necessarily correlated, and mutually condition each other. In this, another proof of my theories is afforded.

Let us proceed still further in our examination of evidence. One of the strongest corroborations of scientific physiognomy that I have received is in the showing of the correlation of the functions of the muscular system with those of the brain and nerves. As Mr. Haeckel has told us, the efforts of the amphibia to accommodate itself to terrestrial life advanced greatly the power and capacity of the muscular system; hence, of the mental powers. Thus it will be seen that these operations are correlated; that, in fact, muscular movements are in themselves *mental* to a degree; not as highly special-ized, it is true, as the faculty for pure abstract reasoning, although I believe this faculty will eventually be proved to have an intimate relation with the muscular and fibroid systems. The increased activity of the muscles necessitates increase in the size of the skeleton or osseous system; also, in the power of the nerves and size of the brain; hence, of mental activity and higher intelligence; for, as the anatomy of the higher animals (dogs and apes, for example) shows that the power of speech is denied them solely on the ground that they have not that development of the larynx, tongue, and lips essential to the quality of speech, which is found among the lowest human races even, and the latter do not possess that perfection of the muscular system which gives the power for perfected speech, such as is found in European races, for example. Speech is thus shown to be a physio-logical gift, as Mr. Haeckel observes—not a "divine" one; that is to say, not in the sense in which that word is com-monly used. I believe every created thing to be divine and emanating from the Creator, whether it be an oyster or an ape, and the reason why neither of them speak is not from a lack of divinity, but because of the absence of a suitable physiological development, each step of which is just as divine one as another; the first step in evolution illustrating the power of the Creator as much as the last one. All are divine, infallible, and unerring.

CHAPTER XI.

SIGNS OF HEALTH AND DISEASE IN THE PHYSIOGNOMY.

"Every disease is a protest of Nature against an active or passive violation of her laws."—OSWALD.

"Unnatural food is the principal cause of human degeneration; it is the oldest vice."—*Ibid.*

The practical value of scientific physiognomy is nowhere more apparent than in the exposition it makes of the construction and conditions of the internal organization of the human body. By the face alone we are able (if we read it scientifically) to distinguish differences in the form, power, and ability of the several visceral organs and systems. Physicians have long understood the value of the pulse as an indicator of health and disease; the tongue, also, discloses both healthy and diseased conditions of the various organs, tissues, and systems hidden from the sight and touch of man; the face, too, has been relied upon to some extent in diagnosing the changes and conditions incident to disease. Yet the face, as an exponent of the form, size, and natural power of the different organs and systems of functions which constitute the organism as a whole, has never been put before the world until now. This knowledge, added to an understanding of the facial signs of health and disease, will be of incalculable advantage, particularly to mothers, and, indeed, to all who are desirous of understanding and conserving their mental and physical powers. I design, in this chapter, to give a few of the prominent signs by which diseased and healthful conditions and organisms can be ascertained.

To a thoughtful and observant person, the face will seem naturally to be the exponent of the entire organism. It has evolved just in the same way that the various organ-systems have developed. From the expressionless faces of the lower animals, the human face has gradually assumed its present degree of perfection of form; as the evolution of the race continues, it is probable that there will be additions and changes of the physiognomy to suit the altered mental and physical conditions which evolution will entail. There has been a constant change and addition of expressions in connection with the evolution of the physical and mental powers. The physiognomies of the most advanced peoples are much more expressive than those of the lowest races; the physical powers of the former are more highly specialized than those of the latter; in fact, we find that the mental powers keep pace with physiological development all along the line of progressive growth.

The forehead, chin, and defined nose are the latest acquisitions to human physiognomy; none of the lower animals possess either; neither have they the same degree of reason, conscientiousness, and mechanical ability as the developed man; and the signs for these faculties are found in the forehead, chin, and nose. None of the lower animals have a nose that rises from the plane of the face, except the nose-ape (*Semnopithicus nascius*), and in this species of animal the projecting nose is not indicative of superior intelligence as one might imagine, although it is far greater in size than the noses of many human beings. The question of quality must always be taken into account as well as form and size. The nose of the nose-ape lacks the right form; that is to say, it is not the same shape which in man would denote superior intelligence; it is long and sharp-pointed, like a gimlet. I regard this peculiar nose as having had its origin in some physical necessity in connection with the getting of food, or that it originated by natural selection, and thus, augmented by inheritance, became a permanent type. Whatever may have been its origin, it is wanting in other essentials in order to be, according to the law of physiognomy, an indication of superior mentality. The quality is wanting; also, the right shape; therefore, this peculiar type does not militate against the statement that "the size of the nose is the measure of power." The noses of all inferior races of men, and of all inferior persons among the superior or cultivated races, are proofs of this law. Observe, for example, the noses of Locke, Newton, Des Cartes, and Leibnitz; theirs are the largest of any historical characters. There is no record in history of any person having attained eminence in any department of life whose nose resembled that of a Bushman of South Africa or that of a Congo negro.

The nose, as has been shown elsewhere, is an indicator of both lungs and heart; and as man depends upon his breathing and circulatory power for his ability to perform almost all of the useful and great acts of life, the importance of the high development of the nose in regard to size and form must be apparent to the reader; therefore, any peculiarity in this feature, which would indicate a deficiency in the action of either the lungs or the heart, would necessarily afford the clue to the grade of mentality of the possessor of such peculiarity. When we wish to discover the natural construction of the lungs and heart, and the power and vigor of the circulation, as well as the ability of the lungs to oxygenate the blood, we must look to the size and shape of the nostrils. and nose. If the nostrils be small, the lungs will be small also; and the heart, not receiving a large supply of well oxygenated blood, will not of course be as powerful as where the supply is greater. The natural or inherited quality of the individual is useful in estimating the strength or weakness of the internal organs, and the power of their functions; this, too, must be taken into account in forming an opinion in regard to their action. The texture, color, and clearness of the skin and eyes, as remarked elsewhere, will assist in arriving at the *quality* of the physiology of the individual.

These characteristics of the natural physiological conditions give us an understanding of the mental powers, for mind is only a question of physiology. Although we have been taught that it is something superior to the body, we know that it cannot exist apart from it, and cannot be regarded as an entity. Now, when we observe a person whose nose lies flat, or nearly so, against the face, we know directly that his mental construction is of a very low order, from lack of the physical assistance which a developed state of the lungs, heart, and stomach renders. A low flat nose denotes a low grade of intellect—low, because there is not the proper apparatus for assimilating enough of the constituents of the atmosphere to give noble aspirations and lofty and vigorous thought. But as my design in this chapter is to treat particularly of the facial signs and indications of health and disease, and of healthful and diseased conditions, both natural and acquired, I shall pass by the meanings of mental significations in the countenance, and confine myself to the purely physiological or pathognomonic aspect of the physiognomy. Physicians in all ages have understood many signs of disease and health as shown by the various expressions and changes of the human face. Hippocrates and Galen, the most ancient medical writers, have left us some opinions in regard to signs of disease which they had observed. Hippocrates tells us that it is a bad symptom "when the eyes of the patient shun the light, when they begin to squint, when one appears smaller than the other, when the white begins to redden, the arteries to grow black, to swell, or to disappear in an extraordinary manner;" and he adds, "The more the posture of the patient approaches that which was habitual to him in a state of health the less the danger."

The natural predisposition to many diseases can be known by the peculiarities of facial construction. With this knowledge once gained, the individual will be able to ward off disease by using such precautionary measures as hygienic law dictates. All hollows in the countenance denote weakness. If these hollows are natural, then the defect is constitutional; if temporary, then they are acquired, and can easily be remedied. A small, narrow, retreating chin, or one which hollows inward near the under lip, discloses constitutional weakness of the kidneys. Hollow cheeks in the lower part show weak digestion, or poor assimilative capacity. Hollow places in front of the ear-opening, where the parotid gland is situated, also exhibit less of assimilative power than where this portion of the face is full. I have observed this gland so emaciated as to form deep wrinkles all over it. This appearance shows that the salivary glands are inactive and small; hence, they cannot secrete and supply as much saliva as is necessary for the perfect insalivation of the food received. A thin, pale, and dry upper lip bears testimony to a weakness in the reproductive system; extreme shortness of the upper lip signifies a tendency to weakness of the spine. Shortness of the septum of the nose so that it is level with the alæ, or wings, or where it is observed to be shorter than the sides, evinces a predisposition to bilious disorders. Dr. Simms gives hollow temples as a sign for a weak liver; my observations confirm this as one sign of a tendency to bilious disorders. A sallow skin, a yellow tinge of the sclerotic, or white of the eyes, are other facial signs of a disordered biliary system.

Narrow or pinched nostrils are evidential of weak lungs. This formation also shows a sluggish arterial circulation. Weakness of the lungs gives other indications of their inability to perform their office properly; a pale bluish cast of the skin, with blue or pale lips and nails, arching of the nails over the fingers, sighing and yawning frequently, shortened respiration, narrow and drooping shoulders, and a flat chest, are all symptoms of an imperfect thoracic system. Disease of the heart is also indicated by a blue skin, fatty cornea of the eye, and red and white spots on the face. I have observed, in severe cases of heart disease, the lips and gums nearly black and the skin as dark as if smeared with ink. Another sign of weak lungs is shown by the hectic flush on the upper part of the cheek, just over the malar bones; this flush is an indication of an abnormal condition of the bowel system, and is observed just where one sign for the bowel system is situated.

The thoracic or lung system is dependent upon the normal action of the bowel system to provide nutrition—to supply the lungs with a sufficient quantity of blood of a suitable quality to keep them in normal action. If the bowels fail to

19

perform their share of work—fail to provide suitable materials for the manufacture of blood—the lungs become impoverished and decay; and the hectic flush denotes an abnormal condition of the bowel system, and that consequently the lungs have not received their right proportion of good blood to supply their necessities.

A hollow or "scooped" nose—that is to say, a nose which is very low at the centre and lies nearly level with the plane of the face—is always accompanied by a weak stomach, or a tendency to such weakness. Hollow, retreating eyes, and depression of the orbits or surrounding parts, disclose a deficiency of power in the muscular system. If the bones of the forehead do not project well out over the eyes, the bony system is comparatively small; that is to say, it is small in proportion to the other systems in the body. If the sign for Weight be small, the individual will not be able to balance himself as well in walking, climbing, dancing, etc., as where it is found largely developed. A hollow in the centre of the forehead announces a weakness of that part of the memory which is devoted to the memory of events, facts, incidents, and biography. Memory has as many parts as there are faculties. One may possess an uncommonly good memory for names and not for dates, or a memory for colors and not for forms, a memory for tune and not for figures or time, or a memory for faces and not for names. Many forms of nervous diseases weaken the general memory. Nervous shocks will sometimes impair the memory for names of things, for nouns and not for adjectives and the other parts of speech, thus proving that memory has almost infinitesimal subdivisions.

Memory is far more complex and minute in its operations than is generally understood. The learned and ingenious Hooke is said, in his speculations, to have estimated "that the mind is capable of containing three thousand one hundred and fifty-five million seven hundred and sixty thousand ideas." Each of these ideas has its own memory, as a matter of course. I think this estimate underrates rather than overrates the divisions and capacity of memory.

The prevalent custom of the almost universal use of tobacco and alcoholic drinks is not only demoralizing the present generation, but is laying the foundation for a large increase of criminal and defective men and women in the next. Wherever we find the renal or kidney system constitutionally defective, we shall be sure to find the moral nature correspondingly weak. The children born of drunkards have often very narrow, retreating chins, the first stage toward idiocy. Many, if not most, idiots show similar formation, and this indicates enfeebled moral perception and power. The reports of the superintendents for several Homes for Inebriates have fallen under my observation; on comparing them, I find that they are unanimously of the opinion expressed by one of them, Dr. Haynes, of San Francisco; viz., that "in chronic cases of alcoholism, there is a general impairment of all the so-called *moral faculties*, and a corresponding increase of the animal instincts and nature." He also adds: "From our own observation, as a general rule, there seems to be a *change* in the very *morale* of the mind. All continuous mental effort soon becomes difficult or impossible; not only are the perceptions blunted, but the intellectual faculties and the reasoning powers are impaired. This tendency, which plays a very important part in the production of premature mental decay, has been attributed chiefly to three causes; viz., hypertrophy of the left ventricle, chronic disease of the kidney, and degeneration of the coats of the cerebral arteries."

In the face of such evidence, I ask my readers what they think will be the mental and moral constitution of the children born of parents who are in the mental, moral, and physical state of degradation above described. An enlightened and civilized community, it seems to me, should take steps to prevent such persons from becoming parents, or, rather, strike at the root of the evil by refusing to license the wholesale manufacture and sale of alcoholic beverages. This effort will probably be postponed until a true civilizing influence is mingled with our governmental affairs—until that portion of our citizens who are not stupefied with alcohol, and are not in the condition described by Dr. Haynes, come to the rescue. Yet many men in this state of moral and mental demoralization become parents, and sit in the councils of state, and are appointed by large majorities of their fellow-citizens to administer upon the life, estates, homes, and children of that other large class of citizens who could and would save the nation from such degradation and injustice if they were allowed the power. I refer to the women citizens of our country. Until that day arrives, we will continue to have children whose physiognomies tell the sad tale of the degrading habits of their progenitors; and nowhere is the sign for moral deficiency more defined than in the lack of a developed chin. This feature is the most distinctive which marks the difference between man and the next lower mammal—the ape—and also marks his superiority in the development of moral sensibility.

In the light of the knowledge which this system of Physiognomy unfolds, how can the reader longer doubt the unity of the mind and body?—how can he longer doubt that they are part and parcel of a chain of causes which cannot, by any manner of metaphysical sophistry, be separated or divided? The mind is the body and the body is the mind. Maudsley says that "when a man is insane, he is insane to the tips of his fingers;" and Emerson tells us that "a man finds room, in the few square inches of his face, for the traits of all his ancestors." Some of my readers may ask, If you make of the body and mind a unit, where, then, is the soul? Do you not believe in a soul?—in a God? To this I must reply, first, that my individual belief would be out of place in a scientific work, which, as its name implies, is devoted to demonstrable fact; and, second, that it is no part of the scientist's work to treat of *faiths* which have only a documentary evidence as a basis, and these very much distorted and changed by mistranslations.

Bacon remarks that "sciences are facts generalized;" and if, in the great world of demonstrable fact and science, there

is not enough evidence of the existence of an almighty and
an overruling Power, I do not know where one may go to
find it. As to the existence, attributes, and locality of the
soul, which many persons believe to be a part of the individual existence, I can only say that I can give no scientific
proof of it. What I *believe* on the subject might not be
evidence; but of one thing I am convinced: that the Power
that created all things has also the power to conserve the
soul, which—since He has not given us the knowledge of its
locality—it must be His especial province to take care of,
and not ours. Let it be our duty to preserve in the highest
state of purity, integrity, and efficiency the body and mind
which He has given into our keeping, and I cannot but
believe that the soul will be correspondingly pure and noble
in His care.

I think scientific physiognomy has demonstrated one thing
above all cavil or doubt; and that is, in order to be moral or
religious (I here use the terms synonymously) one must seek
to secure and perpetuate a healthful and balanced condition
of the physical organs, or fail in this most important department of character.

Some parts of the memory are affected and weakened by
long continued catarrh; other divisions by nervous shocks;
thus we see the importance of keeping the several parts of
the body in repair if we would be mentally qualified to use
our highest powers. The general memory, as I have shown,
may be strengthened, impaired, or wholly obliterated by certain physical conditions. It may be strengthened by a judicious use of it in the following manner: First, by a slow and
deliberate perusal of whatever subject one desires to retain;
afterward, by a careful review each night of the events of
the day, week, or month. A few moments devoted to this
exercise will produce a decided increase in the memorizing
capacity. It may also be strengthened by the use of proper
foods, and the non-use of stimulants in any form. Alcohol,
and malt liquors, tea, coffee, pepper, and too much animal
food, all tend to stimulate the mind; but all exalted and stim-

ulated conditions are sure to bring reaction, and this reaction will produce exhaustion of the nerve forces. Hence, it is apparent that this process called Memory, which, above and beyond all others, has been considered a purely mental function, is dependent for its power and sustenance upon dietetic and stomachic conditions. Another proof of this dependence is given us in the fact that a deficiency of color in the physiognomy, in the skin, hair, and eyes, is evidential of a weak memory. Now, if the stomach were supplied with suitable materials from which the right proportions of color could be extracted, and if the chemical action of the systems of the body which assist in the process of digestion and nutrition were normal, and if the body received sufficient sunlight, the memory would be strengthened, and this "mental" process would be correspondingly improved. The habitual use of tobacco assists not only in changing the color of the complexion, but sometimes almost entirely obliterates the memory of colors, as well as other departments of memory; and this defect is intensified where this habit is hereditary, where grandfather, father, and son have been habituated to the constant use of this terrible poison. Not only is the color sense defective, and sometimes obliterated, but other physical functions and mental faculties lose their normal power and vigor. The functions of secretion and absorption are obstructed by the presence of nicotine (an active principle in tobacco); hence, the tissues are neither purified of their waste particles, nor are they rebuilt, in consequence of the lymphatics failing to perform their office. These glands are affected in such manner by the active poison of tobacco that normal action is impossible. The proof of this position will be better understood when it is shown that the color sense, or memory of colors, is very defective in men— far more so than in women. This arises principally from the fact that men are generally consumers of tobacco, while women seldom make use of it. The percentage of color-blindness in men as compared with the same defect in women is astonishing and almost surpasses belief. Had we not the statistics of eminent and reputable physicians and scientists on this point, it would be incredible.

Now, upon the integrity of the memory of color the lives of thousands of human beings daily depend; as, for example, in comprehending colored signals and lights on steamships and railroad trains; and as these positions are filled exclusively by men, it is apparent that the safety of the traveling community is jeopardized by the use of a narcotic which destroys this most important department of memory. The *facial signs* of this defect are shown in the livid faces and the colorless, lustreless, and yellow hue of the eyes of those who are under the effect of the poison of tobacco. It impedes respiration, and thus decreases lung and arterial circulation; it weakens the digestion; it impairs the reasoning faculties; it unmans the individual, producing a weakness of the moral sense the same as alcohol, and gives rise to timidity and irresolution in principles and practice; and all these defective conditions, when transmitted to posterity, are intensified and increased many degrees. It is one of the greatest obstacles to the march of civilization, inasmuch as society countenances the perpetuation of the race by those who are degraded and vitiated by the use of narcotics. If drunkards and tobacco consumers were prevented from transmitting their defective organisms, the advance of civilization would be most rapid. An enlightened self-interest on the part of governments would seek to prevent such from inflicting their abnormal conditions upon the unborn, for I claim that they have rights which justice should accord; but, as I have elsewhere remarked, men stupefied and besotted are not masters of *themselves*, and should be coerced into regarding the rights of others by the strong arm of the law until such time as they become reasoning beings.

The facial signs of the diseased conditions induced by the use of stimulants are almost too well known to need notice here; but as they are strong and .convincing proof that all bodily or functional conditions are registered in the face, I will state some of them. The reader will have no difficulty in verifying these signs, for they are to be seen in every grade and phase of society. Bloodshot eyes, the white of the eyes turned yellow; full, puffed, and swollen cheeks, particularly of the lower part, near the mouth; puffed appearance under the eyes; sunken eyes, inflamed condition of the entire countenance, but particularly of the cheeks, where the signs for digestion and the bowel system are located, thus disclosing the inflamed and abnormal condition of the digestive apparatus; swollen and purplish colored nose, exhibiting the perversion and blunting of all those fine qualities, the signs of which are located at the end of the nose. Human Nature, Ideality, Sublimity, Hope, Analysis, Constructiveness, are all vitiated, and sometimes wholly obliterated, as we see by the conduct of the drunkard, from long continued use of alcoholic beverages. In the face of these facts, can any one doubt the reliability of the physiognomy as a recorder of bodily conditions. The signs above described show, also, diseased liver, lungs, heart, kidneys, nerves, and brain, and the entire digestive apparatus.

The eye shows many pathognomonic changes. If bloodshot, as is often seen in those who are habitual drunkards, it denotes cerebral and intestinal congestion. Where the whites of the eyes are very yellow, long continued biliary disturbance is indicated. Puffed appearance under the eyes tells of diseased kidneys. A mixed and mottled eye, where spots and specks of yellow, brown, black, and green are found intermingled, invariably denotes scrofulous tendencies, generally pertaining to the reproductive system or the kidneys; usually, both systems are affected where this appearance is observed. Where a large portion of the white of the eye is very perceptible under the retina while the eye is in its natural position and not cast upward, gluttony or inordinate lust is indicated. A sunken appearance of the orbit of the eye announces a deficient Muscular system, as does also a small eye.

The facial sign of healthy and diseased conditions of every feature of the physiognomy has been treated of in this chapter, with the exception of the upper part of the forehead. This part of the face requires no particular investigation as to health and disease. The upper part of the forehead has no movable or soft parts, and it is in those parts where expressions can be observed that diseased conditions are most apparent. The natural formation of the forehead, however, denotes tendencies to healthful thought or to sluggish and stupid action, not only of the brain, but of the functions of the viscera. A forehead, the upper part of which shows a not too abrupt line of inclination from the eyebrows backward, discloses a common sense, mechanical, and rather quick motioned person. This formation accompanies the Osseous and Muscular systems; hence its practical and mechanical ability. This combination of systems indicates quick, active persons both in their mental and physical powers, and this quickness results from an active arterial circulation and strong lungs. In this way we get the clue to the construction of the internal viscera, simply by the outline of the forehead. Comparative Anatomy is infallible in deciding character by form alone, and in this instance, as well as in all the indications in regard to character, we must rely upon comparisons made and proved.

A forehead, the upper part of which is very full and projecting forward and outward from the eyebrows, is evidential of a dreamer, a theorist, a slow, impractical person—one who must be helped by others, or do with little of this world's goods. This formation of the skull belongs, of course, to a body which corresponds in its build to the brain; that is to say, the secretions will all be slow in forming; the lungs, relatively small; the arterial circulation, consequently, not vigorous; and every movement of the body will necessarily be slow and deliberate.

All these differences, and many others, can be predicated by observing just this portion of the face alone, even if the entire body and the rest of the face were shrouded from view. When Physiology and Anatomy are taught thoroughly in our schools and colleges, the amount of useful knowledge

they will render to the public will not be equaled by any
other department of science. These studies, added to scientific Physiognomy, practically applied, would, in two generations, go farther towards regenerating the world than any
system of ethics of which I have knowledge. I hope that
those mothers into whose hands this book may fall will
commence to teach their children the meanings of the forms,
colors, and features of those about them and those with whom
they associate; the localizing part of the science and the
forms, colors, etc., can be taught to children as easily as
geography. The localizing of signs in the face is somewhat
similar to descriptive geography, and far more interesting.
The philosophical, or theoretical, part is for more mature
minds.

If time permits, I shall endeavor to write a primary work
for school-children. I have been encouraged in this project
by the solicitations of many eminent educators. In the
meantime, parents and teachers can draw the attention of
children to the subject, by asking them what they think
is the meaning of certain forms of the nose, for example; and so on, of other features; and then proceed to explain the meaning of natural formations, such as the arch
representing superior power and strength wherever found;
the beak of the birds of prey—the vulture, the condor, etc.—
representing rapacity, love and power for overcoming, desire
to acquire the resources of others, etc. Then take up the
meanings of other formations of the nose—the scooped or
flat nose, representing weakness; then proceed to the indications and meanings of other features and colors. The majority of children can thus be taught by special effort on the
part of parents. My own children have learned a great deal
of Physiognomy from hearing me discuss the science, without
any attempt on my part to teach them; and, when quite
young, could select suitable associates and companions by
this knowledge.

If Physiognomy were taught as a part of our educational
curriculum, our children would be able, when they become

of marriageable age, to select suitable companions for marriage, both as regards physical powers and mental and moral characteristics, and thus be spared the great unhappiness which falls to the lot of many—I might say most—married couples. This is the result, mainly, of being unsuitably mated; this unsuitableness, in most instances, is caused by ignorance of the disposition and of the mental and moral character of each other.

The interests of morality, true religion, and true civilization would be enhanced by the practical application of scientific principles to the reproduction of the race. Persons suitably mated—that is to say, harmoniously united in regard to the right combinations of forms and traits—would insure greater perfection in their children than if the whole matter of reproduction were left to chance and ignorance or inharmonious conditions. I cannot conceive of a nobler ambition in a woman than the desire to be the mother of superior or perfected offspring, but the mother alone cannot achieve this result; the father, as well as the mother, must make himself amenable to righteous law—to hygienic law—if this result would be attained. I believe this ambition will be woman's some time in the future, and, by bearing less children and better ones, true progress will ensue. By this method, humanity and civilization will advance—the real, genuine civilization; not this wretched, barbarous, unjust, immoral condition of society which is with such supreme satisfaction denominated "civilization," but a higher, more just, moral, and truly religious grade of development will evolve in the order which the law of evolution or progressive growth dictates. This law can be assisted in its operation by the co-operation of man—by the exercise of his reason and moral sense; or, it can be retarded by the ignorant and superstitious. The law of evolution can be traced by any observant person who will take time to consider the growth of organized beings and the progress of tribes, races, nations, and peoples, as recorded in animated nature and historical record.

My idea of civilization would be shown in that condition

of humanity which seeks to make the laws of God the great
aim of life.   By this I mean that the laws of Nature should
be practically applied in every department of life, to the
domestic and social relations, to marriage, to hygienic living,
and the reproduction of the race, and in all ways that natural
law can be applied to elevate the human family.   The term
"civilization," applied to the semi-barbarous condition from
which we are slowly, yet surely, emerging, seems like a grim
satire, and would be ludicrous did it not unfold an age of
superstition, ignorance, immorality, injustice, and irreligion.

CHAPTER XII.

HYGIENE.

"Medical science is the art of amusing the patient while Nature performs the
cure."—VOLTAIRE.

"Habits and the use and disuse of organs are certainly of the greatest importance as efficient causes of organic form."—HAECKEL.

The resources of Nature are nowhere manifested in a more
wonderful manner than in the variety and abundance of nutrition which she offers to all organisms, whether of plant,
animal, or human life.   This abundance and variety give opportunity for that natural selection which every kind of
organism makes in sustaining and perpetuating its existence.
The primal elements, in their natural and crude state, could
not maintain man at all, but must pass through many chemical processes before they are suited to his state of existence; but as they pass from one state to another in their
upward march, they serve as nutrition for other and lower
forms of life.   Man could not live on earth, air, and water
alone; yet in these the plant finds its nourishment, although

for its use they must be purified and their essences extracted by the action of natural chemical laws.

After plants have developed, then animals can take into their organisms as food the stalks and leaves, mingled with air and water. Man then uses the more refined parts of both plants and animals, the fruits, roots, nuts, etc., and in many cases must still further refine his foods before they are fit to form part of his body by subjecting them to the fire, in boiling and roasting. Thus every part of the plant and animal serves to keep alive other organisms, and at the last the refuse falls to the ground to vivify the soil, and assist in producing other plants, which feed other animals and men; and thus the circuit of life is kept unbroken, one complete chain, commencing with air, water, and earth, and ending in air, water, and mineral substances—a perfect illustration of eternity.

It is a well established truth that the natural surroundings of man control and create in a great degree his individuality. If these surroundings, in the aggregate acting upon his organism, fashion and shape his mentality—his physical powers, sustain and nourish his blood, bones, nerves, muscles, and tissues of every sort whatsoever, it would seem possible to discover and separate the different influence and elements at work in forming him, and analyze their character and properties. Hence, we must infer, if the origin of these elements can be ascertained, as well as the effects they produce upon man, that the destination of the several elements of creation and sustentation can be traced to the localities which they inhabit in his organism, and the exact parts which they each nourish and rebuild, and the kind of character each is instrumental in creating; also, that as these elements give form, quality, and color, each according to its nature and power, it follows that man's character must be just as susceptible of analysis and comprehension as are the materials and elements which, aggregated, compose his entire individuality. I claim that this can be known. My intention in this chapter is to teach how this can be discovered. A superficial observer or thinker may not succeed as well as one who has given many years to its investigation; still, with the assistance here rendered, a practical person will find his efforts crowned with success.

CARBONACEOUS MATERIALS,

as all physiologists know, supply warmth and heat. They also create fat, which covers most of the organs of the body, invades the several tissues, assists the skin in retaining the heat of the body, and keeps it up to its normal standard of temperature; but where fat is excessive in its development it impedes functional activity, shortens the breath, decreases the action of the heart, prevents rapid locomotion; in short, leads to abnormal action of the motive and circulatory powers. All this induces love of ease, sensuality, carelessness, lack of intellectual vigor, and a preponderance of the *domestic* faculties, which are simply vegetative in their nature. This element, carbon, which in excess creates, induces, and sustains this class of character, is most easily traced from its original source in the atmosphere, by its action on plants, up to animal life; thence onward, after many transmutations, we find it situated in the organism of man, and creating the above mentioned type of physical and mental characteristics. In its normal proportions it adds to the comfort and usefulness of man, giving warmth to his friendly and social nature, heat and force to his physical needs; but, in excess, leading to torpidity and inertia—in short, to abject laziness. This, in brief, is the course of this primal element, and its destination and ultimate character.

NITROGEN.

Nitrogen is very different from carbon in its character and the localities which it affects. This element is found largely represented in our foods. After being incorporated in man's organism, it is found to be the chief constituent of the bones, muscles, nerves, and brains. The phosphates derived from nitrogenous foods create and sustain the bones, nerves, teeth, and assist largely in the structure of the brain. Other parts of nitrogenous elements sustain the muscles, fibres, and fleshy parts of the body. Here we find this elementary principle located, and its character defined and understood. Nitrogen, after being organized in plants and animals, and lastly having become a part of man's organism, is shown to furnish those organs and systems of functions, and consequently those *mental faculties* which are created by an excess of bone, muscle, brain, and nerves. The mental faculties and powers evolved and sustained by these organ-systems are mainly formative; like themselves, they represent size, form, color, and system; hence, they are instrumental in producing art, mechanism, literature, and science.

OXYGEN

is introduced into the system by means of the lungs, skin, and food, and assists in the combustion of those materials which serve to renew the worn out tissues, and also those matters which are to be cast out as waste after having been burned in the body to furnish its heat. It is likewise instrumental in maintaining the normal standard of heat in the system. Where we find an organism that inhales oxygen in excess, we observe certain organs and functions in more active operation than where there is a lack of this element. As oxygen creates heat by combustion, it produces color, and color shows heat and activity; and so we come to the conclusion that an excess of oxygen in the organism of man endows him with activity, warmth (by reason of activity), and color.

The organs involved in the production of these phenomena are the lungs, skin, blood, and tissues generally, for the action of oxygen, where it is excessive, shows its effect in a general and very decided manner. It causes an increased development of the lungs (the inhabitants of mountainous

regions are proof of this), heart, liver, and skin, as well as
an increased activity and quality of the brain and nervous
system. By reason of this excessive action in this class of
organs, an increase in those mental faculties which these
organs create and sustain is induced.

To those who have never thought of relating mental faculties with physical functions, let me ask right here, Whence
do they suppose that these faculties' derive their power?
One may answer, from the brain or nerves. That would be
putting too much work on these already overworked organs.
If the brain or nerves are competent to perform all these
operations, what *need* is there for heart, liver, lungs, muscles,
or intestines? The human body is the most perfect system
of co-operative labor of which it is possible to conceive.
Each part must perform its own work, and each receives its
reward according as the work is well or ill done. If any
part is *badly* performed, *all* suffer, but the chief sinner the
most, which seems to the finite mind a just penalty.

The mental faculties to which an excess of oxygen in the
physical organs gives rise are, first, a high and active quality
of brain and nerves; hope, ambition, cheerfulness, amiability, analytical power in art, mechanism, literature, science,
and discovery; rapidity of thought and motion, purity of
conduct, and elevated and lofty sentiments and desires. All
this will plenty of oxygen give to the human family. It is
most astonishing, in the face of the fact that so little pure
air is inhaled in civilized communities, that people are as
good and pure as we find them. But Mother Nature—mu-
nificent dame!—is forever bringing forth from her capacious
storehouse counterbalancing and remedial agents; endeavoring to bring up her family in such a manner that she can,
some day, be just a little proud of her children; and "when
she will, *she will*—you may depend on 't."

HYDROGEN,

although one of the greatest requisites of life, when found in
excess is a great detriment to the physical and moral equi-

librium of man, and, consequently, *injurious* to his mental and moral powers. The human organism is estimated to contain seventy-five per cent. of water. It enters into every part of man's system; even the enamel of the teeth, the hardest substance in the body, gives out a small quantity of water under chemical analysis. Now, where there is so great a proportion of so unstable a material as water, one can readily reason that an excess of this element would be productive of a very unbalanced condition. It is well known that liquids tend to coldness and inertia, while solids give form, stability, and power. These different modes of action are inherent in the very nature of these two qualities; hence, we are forced to conclude that where there is an excess of hydrogen or water in the system a cold, fluid, inert, incapable condition will be the result. We cannot entirely change the inherent nature of an element by changing its locality. Water will always be moist and shapeless wherever found, and can only assist by chemical action other elements in their architectural or formative efforts. It is entirely chemical in its action, and although it is so essential to life, and to all the operations of organic life, it does not tend directly or positively to form or shape any part of the organism, but acts negatively as an assistant of organic and chemical processes. Thus the character which derives its power from an organism in which there is an excess of hydrogen or water will be negative, cold, calculating, passive; hence, will be disinclined to effort and prone to dishonesty. "For one must live," such reason, and, as they are disinclined to personal effort, they endeavor to subsist on the mental or manual efforts of others. They have the faculty of Calculation very large; this gives worldly planning as well as mathematical power. Gray-eyed people (that is, those who possess the whitish-blue eye) are great calculators. Such persons are cool in their feelings, made so by excess of fluids in the body. The parts of the organism affected by excess of hydrogen are the tissues and the various fluids and humors. The eyes tell this condition very

20

well by the light color of the several humors which are in a fluid state.

By thus tracing, analyzing, and locating the four primal elements or constituents of man's body, we are able to say definitely which system of functions each element sustains, and from this we can tell to a certainty what mental faculties and manifestations they give rise to. The following table gives a condensed statement of the derivation of functions and faculties evolved by the action of the elementary principles of Nature:

CARBON produces inertia, sensuality, domesticity, vegetative.....................................
HYDROGEN produces fluidity, coldness, torpidity, dishonesty....................................... } Chemical.

OXYGEN assists formative efforts, activity, warmth, ambition, purity, semi-mechanical......... { Architectural.

NITROGEN creates forms, art, literature, mechanism, science, mathematics................. { Architectural and Mathematical.

This analysis gives a correct statement of the elements from which human character derives its powers. Thus far has scientific demonstration taken us. It far surpasses in accuracy all of the meandering and meaningless metaphysics of the past ages, and puts to flight as well many theological speculations; for, as Oersted remarks, "The laws of Nature are the thoughts of God," and science is an exposition of these laws set in motion by the Infinite. Therefore, if we read them correctly, we must believe them and abide by their teachings.

The science of the conservation of the vital forces of the body should rank among the first of the world, and colleges for the study of prevention of disease are now more needed than those institutions from which young fledglings called doctors are turned loose upon the world, to "practice" upon the poor victims of disease, who, not understanding the simplest physiological law, have become weakened by disorders which Nature, were she allowed full sway, could remedy. With the advance of civilization many hitherto unknown ailments have appeared which baffle the skill of the most learned physicians, yet their origin is traceable, in nearly every instance, to flagrant violations of well known sanitary laws. The yellow fever has been prevented by scrupulous regard to cleanliness. Cholera originates in filth and by ignoring commonly understood natural laws. Typhoid fever is also a preventable disease, the result of a total disregard of laws known to most persons. Foul air, imperfect drainage, and improper diet are the chief causes which engender this disorder. With this knowledge of their origin, remedy, not prevention, seems to be the favorite method of disposing of these scourges.

If our ministers of the gospel could understand religion as having its source and seat in the organic structure of the body, and would incorporate into their theologies a wise regard for sanitary laws, as Moses did, we might, by years of preaching on the subject, become a righteous and healthy people. Religion is not a theory of belief merely, related to a spirit in some mysterious and incomprehensible manner; but true religion should be, taught as being related first to our physical functions, to our diet, to the air we breathe, to the manner in which we clothe ourselves, and to the scientific reproduction of the race, to the end that the mind may be capable of sound and lofty aspirations, and that our bodies may become "temples of the living God."

The ancient idea of charms, incantations, and fetiches finds its analogue in the minds of many of the modern religionists, who, after preparing a meal, composed, let us say, of sour or heavy white flour bread, fat pork, soda biscuit, and strong coffee, ask a blessing on what has already been cursed in its preparation, by its palpable violation of the laws of digestion and hygiene which God has ordained to be observed in order to produce health. A few words spoken over an ill prepared meal cannot, even by the greatest amount of faith, cause the stomach to assimilate it and give forth healthful juices, to be converted into good, honest bone, brain, and muscle. No, reader—miracles do not take place in Nature; law, settled and defined, is the will automatic in her domain. It is our duty to become skilled readers of this law, and to understand the meanings of the indications which are spread out before us in great abundance, not covered over nor concealed by mysterious devices, but everywhere endeavoring to make their meanings known; and, when we have succeeded in comprehending this law, we will (if we desire to be healthful and religious) put ourselves in intelligent obedience to it, looking to the natural operation of law for beneficent results, and expecting nothing from faith, mystery, or miracle.

THE AIR WE BREATHE.

The air we breathe is the first and most important constituent of our physical and moral well-being. We can exist for a long time on improper food, or even without any; but life cannot be prolonged beyond two minutes without a supply of fresh air, and death is instantaneous under certain atmospheric conditions.

Since the discovery of ozone by Schoenbein, the knowledge of atmospheric laws, together with the powers and properties of the air we breathe, has progressed amazingly. Ozone, which was at first considered to possess health-giving principles, entirely beneficial in their action, has been proved, by experiments on animals by Dr. Richardson and others, to possess the power to produce certain forms of disease where the ozone is in excess of the other constituents of the atmosphere. In one instance mentioned by Dr. Richardson, a room was loaded with this gas for the purpose of experimenting on animals. "In the first place," says the doctor, "all the symptoms of nasal catarrh and irritation of the mucous membranes of the nose, the mouth, and the throat were rapidly induced; then followed free secretions of saliva and profuse action of the skin-perspiration; the breathing was greatly quickened, and the action of the heart increased in proportion. When the animals were suffered to remain yet longer in the room, congestion of the lungs followed, and the disease called by physicians congestive bronchitis was set up. By further experiments, it was shown that these effects were developed more freely in the carnivorous than in the herbivorous animals;" thus showing it to possess greater powers of destruction to men, who are meat-eaters, than to the lower classes of animals, who subsist on vegetable diet. While ozone in excess in the atmosphere will repel cholera and other epidemics, it has the properties of originating the symptoms set forth in the foregoing experiment.

This illustration serves to show us that this agent, like all the other forces of Nature, is both creative and destructive. Two of the chief gods worshiped by the Hindoos, Brahma and Siva, are called by them the Creator and Destroyer. These gods admirably typify the ruling principles throughout the material world. Who shall say that in the dim twilight of the far-off ages these principles and law's of Nature were not better understood than now, and their meanings set before the people under the symbolic form of the deities which represented them? But, as all symbolism and ceremony tend to idolatry, and people soon turn to worship the symbol instead of the principle, it is not singular that the real meaning of these gods was lost in the fast increasing mystery and superstition. The naked truth is so much easier to comprehend than where it is veiled or hidden, especially where it relates to Nature, and can be explained by the operation of her laws.

With this evidence before us of the destructive powers of certain constituents and conditions of the air, and of the deleterious effects of a lack of equilibrium in the atmosphere, we yet know that fresh, pure air is of the first importance to health. Ventilation is almost ignored in our public buildings and dwellings; indeed, many persons are afraid of the two greatest benefactors of the human family—air and water. Proper ventilation should be had in all buildings; it is the first and most important requisite of life, and should be our first consideration. A room is not ventilated unless there is a stream of fresh air moving through it, the means for which can be secured while building. Transom windows should be placed over all bedroom doors, and flues for ventilating should be built in the top and bottom of all other rooms—the lower flues to convey fresh air into the rooms; the upper ones to carry off the impure air. This can be done at slight expense. Such arrangement will prevent many diseases; bronchitis and consumption are both preventable; even when inherited, they can be eradicated in almost all cases. Dr. Marshall Hall says of the latter:

"If I were seriously ill of consumption, I would live out-doors day and night, except in rainy weather or midwinter; then I would sleep in an unplastered log house. Physic has no nutriment; gasping for air cannot cure you; monkey capers in a gymnasium cannot cure, and stimulants cannot cure you. What consumptives want is pure air—not physic; pure air—not medicated air—and plenty of meat and bread."

When we consider that, according to medical statistics, more persons die annually with lung complaints and consumption than with all other diseases combined, it should be an incentive to the greatest efforts toward thorough ventilation. No subject should be allowed to take precedence of this; neither food, bathing, clothing, nor exercise are as important as fresh air. It it the primal necessity of life. If one were to bathe in water in which a dozen persons had previously bathed, he would think that it was a filthy act; yet how many will sit for hours, perfectly content, in a public hall with hundreds of others, inhaling the effluvia from decaying lungs, congested livers, inflamed stomachs (not to mention the odors of whisky and tobacco), and teeth putrid with decay, bathing their lungs in this putrescence with the utmost indifference; a suggestion of uncleanliness never, apparently, crossing their minds.

THE FOOD WE EAT.

In the matter of food, the practices and customs of our civilization are no whit better than our ventilation. Most of the food we partake of is medicated while in process of preparation; yet the majority are as indifferent to this fact as they are to poisonous air; indeed, I think many do not regard as medicines the articles which are mingled with their daily diet. Soda, which is so commonly employed in cookery, is a drug, used sometimes as a remedy and for chemical purposes. It should be kept exclusively for these uses. Soda is found in the human system, it is true, but before entering the body it must become organized; that is to say, it should be taken only as it is found mingled in the juices of plants and in air and water. Taken in this way, it is productive of health. None of the components of the body should enter in their crude state, as in this form they are poisonous and detrimental, with the single exception of salt.

The daily use of soda in food, as well as all forms of "baking powders," injures the mucous lining of the digestive organs, weakening them and inducing various disorders. ˙ I would advise housekeepers to use it only for soap-making; combined with fat, it makes good soap; thus being of use instead of an injury. The fat of hogs should never be put in the human stomach; for, if fat and soda combined make soft-soap, what must be the condition of that stomach in which these two articles are daily mingled? The flesh of hogs is infested with a parasitic worm called *trichina spiralis*, which, if taken into the human system, as it sometimes is with pork, nearly always proves fatal. Pork is the most indigestible of meats, and should never be allowed in the diet of those who desire to live up to health principles. Pork is productive of scrofula and consumption, and no other meat will induce these diseases. Veal is an immature article, and destitute of those juices which make the mature beef so nutritious; it is very indigestible and innutritious, and should be sedulously avoided.

Fine white flour is another article which I believe annually leads thousands to the grave. In its preparation it is divested of all those elements (phosphates) which assist in building brain, the bony structure, the muscles and nerves. Lack of bone produces in the young the disease known as "rickets," as well as bow-legs and early decay of the teeth, and a poor, weak bony system generally. The fineness of the flour causes it to become impacted in the stomach, thus preventing the digestive juices from permeating it; causing dyspepsia, liver complaint, constipation, and other disorders follow in their train. No language is strong enough to condemn the daily use of fine white flour. No mother should attempt to raise her children on any kind except the unbolted wheat or Graham flour. Fine white flour should be relegated to the medicine closet, to be used for poultices, burns, erysipelas, cholera, etc. In cases where the digestion has become impaired by the improper use of soda and other medications, fine white flour bread acts remedially, soothing the injured mucous membranes.

Every mother of a family should provide herself with a good hygienic cookery book, and with its assistance should be able to bring up her family healthfully, and with little trouble.

Black pepper is another substance which should have its place with the drugs of a household, and not be used with food; it is highly indigestible, and inflammatory in its action, in many cases inducing disease of the stomach, liver, and kidneys. It is useful in treating the ailments of fowls, and is a good antiseptic; useful in preserving meats and fish from decay, in place of salt.

Sugar in its concentrated state, as commonly used, is destructive to health and life. I do not think it proper for use except as a remedy in certain wasting and exhausting ailments. As Nature has so judiciously mingled sugar with our foods in fruits and vegetables, we should be content to receive it in that way. A healthy person should religiously abstain from its use in other forms. It is an absurdity to think we can improve upon Nature's fine chemistry, or that we can combine sugar as healthfully as we find it in fruits, etc. Gout, liver complaints, biliousness, headache, constipation, and kidney disorders are often induced and aggravated by a too free use of this condiment. Some mothers

say that their children cannot eat their food unless it is
sweetened. This arises from a cultivated habit. Had they
not been accustomed to its liberal use, they would eat their
food with a greater relish. Where the diet is simple, real
hunger and keen appetite are the result, and there is no necessity to coax the appetite with dainties. Hunger is the
only safe guide.

THE DRINKS WE CONSUME.

If I were imbued with the zeal of Peter the Hermit, I
would go forth upon a crusade against tea and coffee drinking, especially directed against the men and women of America. The peculiarities of the climate of the New World,
added to the nervous tendencies of its inhabitants, combine
to render the free use of these beverages, as now indulged
in, improper and unsafe for a race desiring good health conditions. Men, whose occupations are out of doors, can better stand the demands made upon the system by these stimulants, but they in time come to feel the injurious effects.
For men of sedentary lives with in-door pursuits, for women
and children, they are wholly inadmissible and should be
avoided.

Tea contains two elements which are very injurious; viz.,
tannic acid and a volatile oil. These are the constituents
which produce the symptoms of sleeplessness and nervous-
ness, and contribute to dyspepsia. If the tea be made
quickly, and drunk without long steeping, these effects are
not so observable. It is by allowing the tea to stand long
that these elements are extracted. Aside from the injurious
elements contained in tea and coffee, the application of the
high degree of heat which is used in their preparation for
consumption, and which comes directly in contact with the
mucous membrane of the throat and stomach two or three
times a day, has the effect to produce catarrh, bronchitis,
and dyspepsia, and also weakens the flow of the saliva and
secretions, thereby contributing to indigestion and chronic
dyspepsia. The Americans are a race of dyspeptics and con-

sumptives, and catarrh is common. These disorders are all preventable, and are traceable in most cases to insufficient ventilation or improper diet. Where consumptive tendencies are inherited, they can be counteracted in nearly every case by a solicitous regard for proper air, diet, water, clothing, and exercise.

These subjects are coming to be regarded as sciences, and are receiving the thoughtful attention of those who have a religious regard for the body and mind. The average length of life is greater now than in the preceding centuries. The statistics of the life assurance companies and of the census returns prove this. It is most encouraging to know that what has been said and written of hygiene, and of the laws of life and health, has taken root in the minds of the people and is already bearing fruit. The tendencies of the age are toward a scientific consideration of man, and his nature and wants are being inquired into by anthropologists and scientists generally, with the view of bettering his condition physically, and this of course will lead to his mental and moral uplifting. The stomach is our creator in one sense. Our happiness and welfare depends upon what goes into it; from it are produced the bones, muscles, brain, nerves, and tissues of the organism. It rests with us, therefore, to choose what kind of these constituents we will have. If the stomach of an individual is constantly stored with pork, fine white flour, soda, sugar, spices, tea, and coffee, he cannot expect to be either a very powerful or a very useful man. If, on the other hand, he subsists upon good beef, mutton, fish, eggs, grains, ripe fruits and vegetables, and milk, in the right proportion, and complies with the laws of ventilation, we have a right to expect both health and morality from him; I am convinced he will have all of these, and many more, good qualities.

I cannot too strongly condemn the practice of tea and coffee drinking, for they bring *horrors* in their train differing only in degree from those produced by intoxicating liquors. How many noble women have been made peevish, irritable, and wretched by tea drinking!—how many make the lives of their husbands, children, and friends barely endurable by this habit! I have known scores of cases of women, and and men too, completely unfitted for the prosecution of business by these two stimulants. Persons who have been accustomed for years to their use find it hard to break off the habit, and, after many ineffectual trials, give up the trial and return to a life of misery. To such as desire to give up these drinks without too much suffering, I can prescribe a course which I have found very easy. In the first place, one accustomed to partake of hot drinks with his meals finds the change from hot to cold very hard to endure. I would advise the use for a while, say for two or three weeks, of cocoa, broma, or shells, whichever suits the stomach best. After a few days the craving desire for the stimulating effects of the tea and coffee will cease. During these days give up as much as possible accustomed pursuits, and pass the time out of doors, or in something amusing. After this the desire for these stimulants will be scarcely felt. After a few weeks of the hot substitute, it can be discontinued without much inconvenience. Very little drink should be taken with the meals; if one had not been habituated to it from childhood, it would never be wanted, and the salivary glands would be much stronger by depending entirely upon them for a solvent. In this respect the animals surpass man; they drink either before or after eating. No man who has a proper regard for his cattle will offer them drink at their meals. In this respect he uses better judgment with his horses than he does in the care of his children.

If all inflammatory substances were kept out of the stomach, the demand for liquids would not be so urgent. Very little liquid is needed for health, unless one is engaged in such labors as create great thirst and perspiration, for nearly all the food we eat—fruits, vegetables, grains, fish, and meats—contain a large percentage of water. One reason why we have so many drunkards is that the food commonly used upon our tables is so filled with drugs, spices, pepper, soda, sugar, and other heating ingredients, that an unnatural

thirst is set up, which is not satisfied or allayed with natural
and simple liquids.   The use of mustard and ginger in foods
no doubts leads to drunkenness by inflaming the coats of the
stomach, thus creating an unnatural heat and thirst.   These
articles are *medicines*, not foods.   The effect which is produced by a paste composed of either of them, and applied to
the skin, should be most convincing proof of their unfitness
for daily use with food.   Truly said the wise man, "There is
death in the pot."

Many persons complain that they cannot use milk, and yet
milk is a natural food, and there is no reason why we may
not use it freely if the stomach be kept in proper and normal
condition.   The stomach must be pure in order to assimilate
milk.   This beverage can be used only with certain articles
of food, and especially by those inclined to be bilious, unless
they conform to hygienic law.   That all can use milk is proved
by the fact that all have subsisted on it alone during infancy.
The reason why it did not then disagree with them is because
it was unmixed with the articles which they now use.   Milk,
wheat, and oatmeal are the only articles of food that supply
every part of man's organism with every element for its perfect sustenance.   As milk includes all the elements for man's
healthful development, one may readily reason from this that
it requires very little in combination with milk to sustain a
healthful equilibrium; therefore, if we contemplate using
milk as part of our diet, it is necessary to dispense with
many articles that would be used without it.   Milk taken in
connection with various articles of food will produce violent
sickness, and sometimes death.   One of the celebrated Ravel
family of pantomimists died in consequence of eating a lob-
ster salad washed down with a glass of milk.   In hot countries especially it is dangerous to use milk with shellfish, or
fish of any kind.   Milk and meat together are manifestly
improper, and were forbidden to be used together by that
great sanitary law-giver, Moses, who wrote, "Do not seethe
the kid in the mother's milk."   Milk used with ripe fruits,
grains, or vegetables is most digestible and nourishing.

Most cooked foods should be eaten moderately warm—
never hot. The high temperature injures the mucous lining
of the whole digestive apparatus, and by increasing the flow
of perspiration renders the skin more susceptible to chills
and colds. The habit of eating hot food creates a desire
for liquids, and thus one bad habit leads to another. These
can all be remedied by commencing to check the first one,
and the others will mend themselves. If one ceases to create a demand, the supply will fail. The celebrated French
physician, Desmoulins, while lying on his death-bed, surrounded by his friends, who were lamenting his loss, and
grieving that they should never have so great a physician,
replied to them, "I leave behind me three greater physicians
than I—diet, water, and exercise." He was right; for these
agents are not only remedial, but preventive and health giving. The artificial mode of living at present pursued makes
a change of diet imperative if we would preserve the race
from deterioration. Fifty years ago in this country nearly
every person worked sufficiently to maintain health and
strength; even children had their allotted tasks, which not
only contributed to their bodily health, but served to render
them more self-reliant and responsible in character. Then a
hearty diet of meat could be indulged in by all with impu-
nity. Now the conditions of life have become changed entirely. Sedentary life is the custom of large numbers of our
population, both in cities and country. It is manifest that
the same food which would be necessary in a stirring, active,
out-door life would be burdensome to the system under just
the opposite conditions. With sedentary habits, a light and
nourishing diet, not too hearty, would suggest itself as suitable; on the other hand, for an active out-door laborer, a
diet containing material more highly charged with carbonaceous matter would be demanded. The foods which supply
the greatest amount of nutrition would seem to be the most
appropriate to the more energetic classes of society. Now,
carbon and nitrogen are the most essential elements of our
diet; these are found in greater proportions in oatmeal,

wheat, beans, peas, cheese, and cornmeal, than in a corresponding weight of meats or fish. Unbolted wheat flour contains nitrogen and phosphorus, both highly conducive to the hygienic equilibrium of the body; yet in the sifting process (as is used in fine white flour) some of the most important elements are almost entirely eliminated. Thus it is seen that the unnecessary refining of foods takes from them their most valuable constituents, and that part of the wheat which would assist materially in nourishing the brain, bone, and muscle, is thrown to the animals, and we really consume the waste; that is, the least valuable part of the grain. A mixed diet of meats, fish, grains, vegetables, and fruits, cooked thoroughly, and without medicinal admixtures, as is the universal custom in America, would be the best diet for hard-working people, with milk and eggs added; but for those pursuing an in-door vocation, meats (except in small quantities) should be avoided, as too heating and stimulating to the system, and requiring more work for the digestive apparatus than should be put upon it, thus taxing the nervous system, and causing irritability and sensitiveness of nerves already predisposed to undue activity by too much seclusion from the open air.

Such, in general, are the rules to be observed in diet. Each individual must consult his own system in regard to the quantity of food to be taken. As a rule, far more food is consumed by the majority of people than is required by the necessities of their existence, and thus much is left for the several systems of functions to struggle with. As they are not able to appropriate all that is taken, and not having the power to expel the excess from the body through its natural channels, various disorders, such as fevers, intervene and burn up the waste material which the overloaded body has struggled in vain to dispose of. In this case, fever is a benefactor; fevers, properly treated on a common sense and hygienic plan, and the system not clogged and hindered by drugs, often produce a change for the better in one's health, and the patient rises from a course of fever with redoubled strength and energy.

Most persons have become so accustomed to the use of meat daily that they imagine they would perish were they to discontinue it. Let such reflect that the peasantry of Ireland live almost exclusively on potatoes, oatmeal, and buttermilk, and this, too, in agricultural districts, where their habits are active and out-of-door employment the rule. No healthier class of people come to this country, and yet after a few years they are afflicted with all the complaints common to our own people. This results from their adopting our forms of living, particularly the daily use of meat, which many of them have told me they could obtain only at Christmas and Easter, twice a year. Their simple diet of potatoes, oatmeal, and buttermilk is found sufficient to sustain life, with the most robust health under severe toil, and to assist materially in longevity. If this be the case, as it indisputably is, cannot our sedentary population become stronger and more healthful by refraining from the daily use of a flesh diet? It is sometimes tried by those who feel the need of a radical change in their health conditions. At first, a feeling of languor supervenes, and the experiment is discontinued because it is thought that the sensation of lassitude is the result of insufficient nourishment. This is not the case. Meat, like many other articles used commonly, is stimulating, and in dispensing with its use one feels at first the lack of the accustomed stimulus, just as the tea-drinker or the dram-drinker does, although not to such an extent. The trial is not continued long enough, perhaps, or it may be that the best substitutes do not replace the meat, such as unbolted flour, milk, and fruit; hence, weakness is the result and the experiment discontinued, and the hygienic system of living is pronounced impracticable and detrimental.

I visited, one summer, a farmer's family in Connecticut, where the daily diet the year round was composed of salt pork, boiled or fried, dried or salted beef, fine flour bread, saleratus biscuit, cakes, pies, pickles, and preserves, with tea and coffee (with deleterious mixtures) three times a day, a few vegetables, and not many fruits; they did not think them worth the trouble of raising. The mistress of the house, a pious lady of about fifty years of age, was troubled every three or four weeks with a bilious headache, so severe as to confine her to her bed, in a state of partial blindness, for several days at a time. She believed it was the "will of God" that she should be so afflicted, and therefore wholly incurable. She stated that she had taken dozens of boxes of pills for it without relief; that her mother had had it before her, and she expected that it would result in death. She was also dyspeptic, and troubled with eructation of gas, great indigestion, and weakness of the entire system. She had, in consequence, become so impoverished by non-assimilation of food that it had led to hemorrhage of the lungs. I saw immediately that her diet was the cause of all her sufferings; I explained my ideas to her, and showed her that she could be easily cured by food alone; and, as I could not submit to her style of living, fresh meats were ordered from the town, and fruits, also. The tea and coffee, pork, pies, cakes, salt beef, and preserves were discontinued, and soups, fresh meats, fruits, vegetables, unbolted wheat flour, milk (which she could not previously drink), and eggs, were substituted. In two weeks the indigestion had ceased, the strength improved, hope was restored (owing to the renewed activity of the liver); the headaches which she had suffered from for forty years disappeared after a month's change of diet, and did not return during the summer, while she continued this course of living; but, after I left for home, the family fell back into the old routine, and I learned that all her former symptoms had returned.

Now, here was a family possessed of all the essentials of hygienic living, except fruits, which they might have bought or cultivated, committing slow suicide—leading a life of suffering, wholly unnecessary and entirely preventable. There was not one of the family in perfect health, notwithstanding that they had the purest atmosphere, plenty of out-door exercise, and were highly religious—that is to say, theological; for I cannot consider any one religious, in its natural and highest sense, who lives in defiance of the laws of God as revealed to us by the laws of digestion and hygiene. Instead of reading good books on these subjects, and applying their laws to make themselves moral, healthy, and truly religious, they passed their leisure in praying over most wretched food, believing that that course would surely "bless" the food to them—food already cursed by its unfitness for human stomachs. To such a pitch of demoralization had generations of this sort of diet brought this family, that not only had physical degeneracy been the result, but in several instances the children of this woman had become so thoroughly corrupt that one died of delirium tremens, one of a nameless disorder, another practiced self-abuse until idiocy supervened, and the father died of drunkenness.

Now, these people had been devout Christians for generations; there had always been prayer in their home, and at church they had been regular attendants. They lived remote from a city with its excitement and bad influences. Yet, as I have shown, the improper diet led to unnatural thirst, and the system craving and consuming intoxicating liquors in large quantities, led to heated and abnormal condition of the reproductive system; and thus one abuse led to another, until this praying, religious family became as low and demoralized as those who had never heard prayer, preaching, or the name of God. There are thousands of just such Christians, who lose the principle and observe only the symbol or form. I often thought, as I listened to the well-meaning minister who, every Sunday, expounded the "word of God" to these people, and who dwelt so earnestly upon the terrible condition of the lost tribes of Israel, if he had dwelt as untiringly on the lost condition of the perishing sons and daughters of Connecticut, he would have succeeded in building up a religion which would have brought regeneration down to a scientific demonstration, and the devil, even, whom he preached about, might have assumed a fairer hue had he been seen through eyes not jaundiced with bile or weakened by dyspepsia.

21

Our country is peopled with families who, with rare exceptions, have the means of providing themselves with a hygienic diet and surroundings. Many farmers' families, as in the instance above noted, do not use their buttermilk, but throw it to the hogs, and the bran of their wheat to the hogs, horses, and cattle, thus depriving themselves of the best elements of their productions. The buttermilk has, as one of its constituents, lactic acid, which is highly conducive to health and longevity. The bran contains the phosphates by which the brain, nerves, and muscles are sustained. This course is not only destructive to health, but expensive, esspecially if for the buttermilk tea and coffee are substituted, and sugar, spices, and other costly condiments are used in place of the bran of the flour. These do not yield nourishment, but assist in clogging the system, thus inducing various diseases, rendering the services of a physician necessary. This, with the loss of business which sickness entails, together with the numerous incidental expenses consequent on illness, and, above all, the suffering of body and mind, all combined, ought to furnish us a lesson which should teach us the value of strict obedience to the laws of hygiene, which are the laws of God. A life-long and daily study of scientific works should be the aim of all; a single chapter read night and morning would, in a short time, add much to one's knowledge on subjects of the first importance to us. Works devoted to an analysis of foods and drinks—also, to the components of the several parts of the body, and of different kinds of nourishment requisite to sustain the operations of the several systems within the body, and which assist its normal action—should form part of the library of every family desiring to lead godly or religious lives.

HOW TO CREATE AND MAINTAIN EQUILIBRIUM IN THE SEVERAL FUNCTIONS.

This system of Physiognomy simplifies very materially the process of building the body upon scientific principles. In the first chapters of this work, a description of the systems which indicate the division of the human body into the Chemical, Architectural, and Mathematical, was given, and it was there shown which distinct powers each of these divisions includes. So by learning what elements of food are required for the sustenance of each part of this division, it is easily seen that this simple and scientific arrangement provides a way by which, through the harmonious and grand correspondence of Nature's laws, the form of the body can be under the almost complete control of the individual, and that any deficiency in either division of the organism can be remedied by using those elements which chemical analysis has taught us are the sustainers of the several systems included in the human body.

The four ruling elements found in all organisms and plants designed for the food of both man and animals are found to be composed, first, of nitrates, or those elements which tend directly to the maintenance of the muscles and fleshy membranes; second, the phosphates, which sustain the bones, brain, and nerves; third, the carbonates, which produce animal heat and force; and fourth, water, which acts as a diluent, and conveys the several materials to the various parts of the body, and forms three-fourths of the weight of the body.

The following estimate of the proportions of food suited to the daily requirements of the laboring man is given by Dr. Albert J. Bellows in his work entitled "Philosophy of Eating," which I advise all my readers to peruse. He remarks:

"The daily requirements are five ounces of solid nitrates for the muscles, twenty to twenty-two ounces of the carbonates for animal heat, two or three per cent. of phosphates for bones and for nervous power, with waste and water to give it bulk, and acids to eliminate effete matter from the blood through the liver, and this food must be so prepared and cooked as to be eaten with a relish, and not too easily digested."

These proportions of the four essential elements are based on the proportions of the same elements found in wheat;

that is to say, in the whole grain; for wheat is acknowledged by all chemists to be the model food among solids, and milk as the model among fluid foods. In order to keep in repair the muscles and fleshy membranes of the body, foods containing an excess of nitrates must be used. These foods are wheat, oatmeal, cheese, lean meats, beans, peas, southern corn, fish, green vegetables, and fruits. Where carbonaceous food is needed—that is to say, where it is found necessary to assist respiration and animal heat or warmth, such as fat furnishes to the system—then a diet composed of those articles in which starch, sugar, fat, and oils abound, must be used. The carbonaceous elements predominate in buckwheat, sweet potatoes, Irish potatoes, carrots, beets, parsnips, all fat meats, butter, lard, sugar, fat pork, fine white flour, and fishes abounding in oil. The phosphates needed for the bones, brain, and nerves are found in unbolted wheat, oatmeal, and in those fishes which have the least fat, in beans, peas, lean meat, and southern corn, and in acid fruits and succulent vegetables.

HOW TO MAINTAIN EQUILIBRIUM OF THE BODY AND MIND.

To maintain equilibrium in the three divisions of the human system, where inherited quality has created a healthy balance of the various powers, a judicious mingling of these elements is necessary. Dr. Bellows gives the proportions for healthful equilibrium as follows, based on the component proportions of wheat, the model food. He says:

"It will be seen that in ordinary circumstances of temperature, muscular and mental exercise, etc., the proportions required are about fifteen per cent. of the nitrates, or muscle-making elements, to sixty-five to seventy per cent. of the carbonates, or heat-producing elements, and two or three per cent. of phosphates, or food for brain and nerves, or a little more than four times as much carbonaceous food as nitrogenous, and seventeen or eighteen per cent. of waste and of water."

The waste part contained in the bran of wheat is needed to assist in digestion; also, the waste in fruits and vegetables— that is, the fibres.

These proportions must be changed to suit inharmonious developments of the body. Where the bony system is weak, or where the brain is too much exercised, then a larger proportion of foods containing nitrates and phosphates must be added to the diet. If, on the other hand, more fat or animal warmth is required to sustain the heat of the body, and to round out and cover up the bony system, then more food containing the carbonaceous elements will be demanded. If the liver is not sufficiently active, then acid fruits must be used in excess, and fats, oils, sugar, and fine white flour be entirely dispensed with.

Thus, by a few simple changes in the articles of food, we can create great changes and improvements in our physical and moral natures. I do not say that a fine mechanic can be formed from one in whom the chemical or vegetative system predominates, or that a genius can be created by an excess of phosphatic food; but a change of diet will modify this system and form very materially, as well as all other forms, and the health and usefulness can be enhanced and longer life insured by accommodating the food to the conditions required. The generations following will also reap the benefit.

HOW TO REDUCE THE VEGETATIVE SYSTEM.

Where predisposition to accumulate fat has been strengthened by generations of cultivation, very great effort will be needed to change the condition. This can be done by a persistent course of diet without weakening the system; on the contrary, it will be strengthened by thus doing. William Banting, a corpulent Englishman, has published a pamphlet in which he gives directions how to reduce the size without impairing the strength or health. He reduced his own weight from two hundred and two to one hundred and fifty-six pounds by abstaining from extra heavy carbonaceous foods and from

the use of milk, sugar, beer, potatoes, and heavy wines,
using only beef, mutton, poultry, veal, and all vegetables except potatoes, beets, and parsnips. He used claret and
sherry wines for drink. I would advise abstinence from
liquids and liquid foods, in order to reduce the bulk of the
Vegetative system. Those possessing this system and form
in excess will gain bulk on a liquid diet alone. An exclusive
diet of raw fruits, except apples and bananas, will reduce
bulk rapidly.

HOW SELFISHNESS, IMMORALITY, AND DISHONESTY ARE INDUCED.

Where the lungs are relatively small the blood will be in-
sufficiently oxygenated in the process of respiration, and, as
a consequence, the fat of the blood will be deposited in the
cellular tissues. If the liver become overburdened or inactive, or the skin allowed to become defective in its action
through lack of bathing, or in any way, then fat will be accumulated. The nostrils of almost all excessively fat persons are very small, and such should live much out of doors,
and take exercise calculated to increase the size and activity
of the Thoracic system. The constant use of alcoholic stimulants impedes respiration, and thus causes fatty depositions.
Too much sleep will produce the same result; also, gluttony
and inaction of the Muscular system. The too free use of
carbonaceous foods induces an inert condition of the entire
organism, as well as an *immoral organization*, for an excess of
carbon is the dominating element in the Vegetative system.
This system, as has been seen, exhibits far less of moral
strength and power to resist immoral tendencies and temptations than either of the other four systems; viz., the Thoracic, Muscular, Osseous, and Brain systems.

I cannot but condemn in the strongest terms the universally wholesale use of *sugar*, which is almost pure carbon.
To its too free use I ascribe much of the present state of dishonesty, immorality, and lax principles so prevalent in every

grade of society. No form of *belief* can neutralize the effect of this condiment as it becomes incorporated into the body and mind, for it cannot affect one without affecting the other. No intellectual process or belief in any dogma can cast it out, or prevent it from influencing the morals according to the law of Nature which punishes excess in every form, in spite of our *opinions* on any subject.

It is excess which produces unhealthful conditions. A man may become so muscular by training as to induce serious disorders. Nearly all athletes and acrobats are short-lived, because they develop one set of functions at the expense of the others. Thus of every system composing the body. It is equilibrium, or a balanced condition, which assists Nature in doing her perfect work. It is excess which hinders her and thwarts her beneficent operations.

WHAT REGIMEN STRENGTHENS THE BRAIN AND NERVES.

The brain can be strengthened by using an excess of phosphatic foods. These foods will not create mental giants, but will give the brain more power than carbonaceous foods. The circumstance of inherited quality is a substantial and an abiding condition, and in order to have talent or genius it must have been inherited. Afterward, talent, as well as physical qualities and conditions, may be modified, strengthened, or weakened, according to the conditions of life surrounding the individual. In these changes and modifications food is a most potent factor.

Another important influence upon health, and inducing longevity, is mental employment. A due degree of mental application assists in maintaining healthful equilibrium of the entire organism. Many of our most distinguished *litterateurs*, both male and female, have lived to an advanced age, and some of them were most industrious thinkers and writers up to the time of their demise. Mental labor lifts the weight, if I may so express it, from the body, and rests the purely physical powers. Every person should give some time to brain culture and expansion, if health and happiness be desired. The sad effects of physical drudgery, unrelieved by mental relaxation, are shown by the statistics of lunatic asylums, from which we learn that farmers and farmers' wives form a larger percentage of the inmates than any other class of society, the wives of farmers showing a larger percentage than any other. The reason is found in the monotony of their lives, which are far more secluded than those of their husbands, with but little out-door occupation to aid health, and little mental occupation. In California, a large proportion of sheep-herders become insane; the monotony of their existence is intensified by a total lack of all relaxation, either social or mental. Their stolid, unintelligent faces are proofs of the relations existing between mental conditions and the physiognomy.

EFFECTS OF ALCOHOL UPON THE BODY.

I have written nothing in this chapter on the injurious effects of alcoholic beverages. Tabulated statements of the effects of alcohol upon the human mind and body are unnecessary here. The sorrow, destitution, depravity, and physical demoralization it causes stare us constantly in the face. Every family in the land suffers in some way from its blighting power. No one is ignorant of the effect of this gigantic curse. Warnings against the habitual use of alcohol have never been wanting since the first drunkard left his orphans to the cold charity of the world. Words have accomplished little toward abolishing the use of the greatest evil in our land. Nothing short of combined action through legislation will ever reach and put down this monster. This action must come from those who are not under the power of its demoralizing effects, whose brains are not confused by its operation, whose moral sense is not perverted by its use. This action must come through the personal efforts of those who are the chief sufferers by its legalized sale—I mean pure

women. Men who are permeated with its effects are not
capable of either resisting its power or of assisting in its
abolition, and we need not look to that part of society for
moral reform. Woman is destined to be the saviour of the
race. Her opportunity will come, and she will use it grandly,
nobly; and the demon Alcohol will some day be throttled by
the delicate fingers of the women of the land by the help of
the little silent ballot.

CHAPTER XIII.

HEREDITY.

"A man finds room in the few square inches of his face for all the traits of
his ancestors."—EMERSON.

"Every birth is a hygienic regeneration. The constitutional defects which de-
generate parents transmit to their offspring are modified by the bequest of an older
world."—OSWALD.

"Ye cannot gather grapes from thorns nor figs from this-
tles," said a wise and observant man long ago. A correct interpretation of Nature will prove that like begets like. We
should have as great a variety of fruits as we have of human
beings if we as constantly grafted new scions upon young
trees that were as diverse in character as the men and women
who intermarry. The law of inherited quality and character
is perhaps the most difficult and complex part of propagation
to investigate and prove. Stock-breeding on scientific principles, as it has been and is now being practiced on an extended and intelligent scale, will go far toward enlightening
us as to the methods employed and the laws observed in improving the various races of animals which are under cultivation.

These same laws put in operation will produce similar results in mankind, for man is an animal not many degrees

removed from the next lower mammal; and when humanity is sufficiently developed in wisdom, justice, and true religion, we shall be able to improve the race upon scientific principles, and create higher types of men and women on a basis which shall bring forth more perfect specimens, and more satisfactory results, than are now produced under the *instinctive* method, which is the only one employed by man and animals, except in those cases where men use their *reason* in the selection of animals for the purpose of improving them by a judicious mingling of forms and faculties.

The inherited nature of man is a more potent factor in his life than all the education he can possibly receive. If an individual is born of a long line of ancestry who were moral and intelligent on both sides, the probabilities are that he will partake of their nature. If a child is born of a race who have lived by lying and thieving, the chances are that he will lie and steal with facility and ingenuity. Parents often wonder why their children are like neither of them, thinking that all children must resemble one or the other of their parents. This ignorance is a proof of how little they have thought on the subject of the reproduction of the race, and shows also that they do not understand the simplest law or principle in connection therewith. While a man will make great efforts to trace the pedigree of a horse which he is about to purchase, and insist that it shall be free from vices and bad blood, he will at the same time take a woman in marriage without even inquiring whether her parents or grandparents were insane, scrofulous, epileptic, consumptive, or idiotic. If the results of his marriage should prove disastrous, and a family of foolish, scrofulous, or vicious children make their appearance, he will, if he be a religious man, attribute this dire calamity to "the will of God," and state that for a "wise and inscrutable purpose" He has been pleased to visit him with this affliction, never for a moment endeavoring to trace these effects to their cause. Not so would this same man reason if the *horse* turned out contrary to expectations. He would accuse the dealer of dishonesty, and would know that the pedigree of the horse was not as represented, and that its parentage was vicious and of bad blood. So blinded by superstition and ignorance are many men that they cannot understand the laws of scientific breeding in regard to the rearing of children, yet are willing to admit their influence in the breeding of cattle.

Ignorance and selfishness are constantly reproducing their own types, and to that extent do these two traits enter into marriage that I am surprised at the number of decent people there are in existence. Ignorance is in some sort an excuse, but in these days of general scientific knowledge some individuals are responsible for the miserable failures in the form of children which one finds in every community. Selfishness must be the ruling motive, since men knowingly perpetuate vicious and sickly types, simply to gratify themselves in the posession of a certain woman, or for mercenary motives. I am informed that a certain honorable senator of one of our Western States had born to him several children by an insane wife after she had been pronounced incurable. I fail to see "the will of God" in such conduct. It is a defiance of God's laws—therefore of His will. It is an outrage on the innocent victims of man's selfishness, and on a community, which, if it were sufficiently imbued with a knowledge of natural law, ordinary kindness, or proper sense of justice, would make laws preventing improper types from perpetuating their abnormal organisms, thus protecting the State from expense in building hospitals, jails, and court-houses, which have to be used in a large measure for the care of the moral monstrosities born of such abnormal parentage. Instead of making reproduction a subject of scientific investigation and preparation, it is left to instinctive action, as with the beasts of the field, or, worse still, to the abnormal operation of the physically depraved organisms with which society abounds. An enlightened and conscientious person would endeavor to select for a companion one whose qualities of mind and body (in combination with his own) would assist in the elevation and perfection of the race, thus obviating the necessity of maintaining the vast armies of professional tinkers which abound in every community. Is a child born diseased in his physical nature?—straightway a medical tinker is hired to endeavor to patch up the defect. Is one born mentally weak?—he is given in charge of a teacher of imbeciles. Is he born morally unbalanced?—then a theological tinker is called upon to try to remedy by "regeneration" the "sins of the fathers which are visited upon the children to the third and fourth generations."

Such a state of depravity ought not in the present general diffusion of knowledge to exist. Society as now constructed is on the most expensive and self-destructive plan that it is possible to conceive, because cultivated and inherited selfishness is the basis of it; and this trait is fostered and increased by the vicious public opinion which teaches virtually that a man is better in proportion to the amount of money he can get, and this leads men and women to marry for money without regard to their fitness for parentage. It leads also to dense ignorance; for, all of the powers of mind and body being bent to acquisition, the mind is drawn away from the study of the natural sciences, and men and women are hence very ignorant of themselves and of the laws which govern their being. This knowledge is the very first step toward a religious life, for no one can be religious while ignorant of the laws of mind and body, and of those other matters which influence greatly mental and physical well-being. The majority of people are far more ignorant of their own physiques than they are of the nature and laws of commerce or of fashionable amusements. They will pass years in learnig some unimportant accomplishment, while they never give one hour in studying how to improve their progeny by superior methods of life.

There are some who think that it is only necessary that the parents love each other in order to create higher types. It is true that this is one very excellent condition in child rearing, but it will not take the place of intelligent, scientific methods; it will not eradicate consumption from either parent; it will not change the scrofulous taint to purity, nor will it alone produce more beautiful or stronger types; it will not give us more moral or beautiful offspring. Nothing but an intelligent, scientific understanding of the body and powers of the mind, as taught by Physiognomy, Physiology, and Hygiene, can improve the race. Blind love nor blind lust will neither accomplish great results. "Cupid should be painted with eyes, and not, as now and as with the ancients, blind. Give love eyes; the scientific Cupid must have eyes. As religion without reason is superstition, so love without reason is lust." Love as a factor in the reproduction of the race at present is a blind force; applied with science it will become an intelligent power.

The woman who consents to become the means of perpetuating a race of drunkards by living with a drunken husband is guilty of gross wickedness, for physical sins are no whit less sinful than moral ones, and lead as I have shown directly to them. Oh, that some latter-day Moses would arise and lead the children of this age to the Land of Promise, flowing with the milk and honey of scientific religion, which is the fulfilling of the law! Religion is not, as many believe, a seventh-day duty, but, rightly understood, is something to live by, to incorporate into our every-day life, and into all the acts of life, particularly our physical life, for it is here that religion commences. No one can be religious with a disordered nervous system, a swollen liver, a dyspeptic stomach, or a scrofulous body, for all these disorders lead to an unnatural and perverted understanding of life and its duties, and cause incompetency in carrying them out. There is no doubt that the doctrine of "infant damnation" was the result of liver complaint and dyspepsia. I believe that many of the horrible and unnatural dogmas which have been foisted upon the world as religions are traceable to men whose bodily conditions had so perverted and demoralized their mental vision as to result in those mental and spiritual monstrosities which the scientific mind of this age is rejecting with disgust. This is the first effect of what little physiological

knowledge there is prevalent in the world to-day.   As these
and kindred truths are spread, we may expect greater results.
An enlightened self-interest will prove to mankind that in
its highest consideration, as well as in its most mercenary
view, a race of children which are the product of selected
conditions are more desirable than those who are the result
of mere instinctive propagation.

Now, the mother is in one sense the architect of her child;
if the mother is a weak-minded, small-nosed, small-souled
being, is it reasonable to expect that she can perpetuate a
race of mental or moral giants? Ignorant men choose weak-
minded women for wives, because they admire their helpless-
ness; it is "such an interesting trait!—it makes a woman
appear so lovely and soft in her manners!" Such women
produce very soft specimens of children. These men never
look beyond their noses; had they a sense of real true self-
interest (not to speak of any higher motive), they would
choose 'for wives the most efficient, capable, and intellectual
women to be found, to be the mother of their offspring; and
the chances are that they would find a fortune in each child
born to them. Men who do not thus choose are very short-
sighted or very selfish, and wish to gratify their own small
tastes or passions by the possession of a woman who will be
good for nothing but a plaything, and of no use for any of
the serious or noble purposes of life. It is related of Talleyrand, a celebrated statesman of France, that, having chosen
for a wife a woman of very inferior intellect, was asked why
he chose such a one; he replied, "She rests me." A high
and noble motive, truly, for a great man!

I have observed that many of the so-called great men of
the world sometimes do very small things. Napoleon is an
instance of the pettiness of a great man, aptly shown by the
memoirs of Madame de Remusat. Lyman Beecher had three
wives, all women of superior vigor of mind; the result was
several uncommonly talented children, and all the rest above
mediocrity. The third generation of Beechers does not seem
to be of the same calibre; the reason is obvious.

Many men seem afraid of women of decided mentality, who are self-reliant and of vigorous intellect. Yet these women would transmit like qualities and bless the world with men and women of stamina and power. Genius and talent are not created in one generation, but where they are exhibited it is the result of several generations of cultivation, and therefore likely to be transmitted by inheritance if the conditions are favorable; and no more favorable conditions can be found than where both parents possess superior inherited gifts. If these conditions cannot be had, then a superior mother is most likely to be the progenitor of improved types. History abounds with the records of great men whose mothers were superior, but the noted children of great fathers are far more rare. Vigorous bodies and strong minds in both parents are highly desirable. For this reason, the conditions surrounding the mother during pregnancy should be the best that can be procured; comforts, with as much ease as is consistent with health, added to perfect freedom for the mother from all selfish and ignorant laws and desires, should be hers. In the domain of maternity woman must reign supreme, or her mission is in vain. Free men are not born of slave mothers; neither petty tyranny nor ignorant assumption should control the dawn of life.

The treatment and education of the child is of far more importance for the nine months preceding its birth than the most expensive curriculum for twice that number of years after being ushered into its next stage of existence. Men who are ignorant of the simplest physiological law concerning women's nature often undertake the control of their wives during the most important period of their life, sometimes inflicting irreparable injury on their offspring by the exercise of an ignorant and domineering will, instead of bringing to bear wise, intelligent law, coupled with tender love and absolute justice.

Another impediment to the improvement of the race is mock-modesty. The prurient shamefacedness of those persons who have held the laws of Nature to be immodest, and have built up, by their example, a vicious because unscientific public opinion, is receiving a shock which will bring them to a sense of their utter misappreciation and misapplication of their opportunities for elevating and improving the human race, which a kind Providence has in times past bestowed upon them. The present general dissemination of scientific truths in regard to the procreation of the race shocks the mock-modest person, but, at the same time, it is doing much to improve public opinion, and is teaching great principles of life which were in former ages not understood, and would not have been tolerated any more than were Galileo's theories of the heavenly bodies. To have allowed these truths would have been to take from the Church some of its power by depriving it of some of its mystery; teaching a knowledge of this all-important science would have seemed to the devout believer in superstition a desecration and a sacrilege; that men had power to improve the race by design would have seemed a monstrous heresy; accidental or lustful procreation appeared to these pious souls the only right method.

Another vital hinderance to improving offspring is the false position in which woman is placed by the decrees of society and governments—a resultant of ignorance and barbarism. Her true position in the scale of creation shall now be explained on scientific principles.

All naturalists are agreed upon the proposition, that the greater number of functions an organism possesses, the higher its rank in the scale of creation. This law is acknowledged by Humboldt, Cuvier, Buffon, St. Hilaire, and all other writers on natural history, anthropology, and natural classification; yet, with singular unanimity, which I cannot believe to be the result of prejudice, they have omitted to make its application extend beyond the animal kingdom. I suppose the reason is that, like Columbus and the egg, "they had not thought of it." A horse is higher than a snail or a snake, simply because it has more functions than either. Woman, possessing *two more functions than man*—viz., those of gestation and lactation—stands, in respect to function, man's superior; yet, with this natural superiority conceded by the law of natural classification, she holds a position inferior to his.

This subject condition of woman is a relic of barbarism— of the age of force, when one being was accounted superior by the power of muscle alone. Now, if it be true (and I think I have proved that it is so) that functions and faculties are correlated, and that for every function there is a corresponding faculty, what are we to deduce, I ask, from the fact that woman's organism comprises two more faculties than man's? To make this fact consistent with other facts of Physiognomy and Physiology, we must conclude that there are two extra faculties related to these two extra functions; and yet this very superiority of woman's organization is often adduced to prove her inferiority to man. If man had two functions in any way analogous to these, the position of the two might be considered equal; but when we reflect upon the comparatively insignificant part man plays in reproduction, the inferiority of his organism must be apparent to any one not prejudiced by venerable mythologies.

I might instance the circumstance of the birth of Christ as showing that man's co-operation in reproduction is not always essential. This instance (and there are several similar ones vouched for in the ancient Oriental traditions and religions) might be referable to a law which operates in some low organisms. This method of reproduction, which is called *parthenogenesis* (virgin generation), may have taken place under the law of atavism, or taking back; for, as is well known, and as I shall show later in this chapter, the law of atavism has apparently no definite limit as to duration of action; for I have, in my own family, one who has rudimentary gill-openings, in this instance placed just in front of and above the ear-opening, instead of back of the ear, as is often the case, and where it would seem that it should appear. Now, this peculiar formation (which allowed respiration through gill-openings) is traced to the primitive ver-

22

tebrates of the Silurian period, millions of years ago. The only way that the birth of Christ can be accounted for on scientific principles is by explaining it as under the operation of these two well known laws. I am not in favor of taking any theological work as a text-book of science, but it seems singular that those who do so should not have thoughts of this method of accounting for what is termed the "miraculous conception of Christ."

Where all can be explained by law intelligently, why endeavor to explain by unintelligible miracle? If the laws of parthenogenesis can operate in one instance to reproduce phenomena which originated millions of years ago, and the laws of atavism can reproduce types of the rudimentary organs dating from the Silurian period, as in one instance which has come under my observation (and naturalists are cognizant of many similar appearances under this law), it is logical and scientific to infer that these laws can operate in other instances. I am not now endeavoring to explain miracles, because I cannot allow that such things take place in Nature; for whatever occurs must be subject to natural law. It is far better for intellectual advancement, and for the interest of true religion and morality, that we should seek to explain all phenomena by the operations of the laws of God, instead of confusing the intellect with a belief in so-called "supernatural occurrences." All naturalists will accept and act upon the following idea of Mr. Haeckel, who remarks: "We are obliged to draw our conclusions according to the laws of induction, in every case in which we are unable to establish the truth of Nature immediately, by the infallible method of direct measurement, or mathematical calculation." And, in this reference to miraculous phenomena, I have endeavored to divest the birth of Christ of all supernaturality, and show that it was *possible* under well known *laws* of God.

Let us return from this digression, and consider whether the so-called intellectual inferiority of woman arises from congenital and natural conformation of the mind, or whether it be not a result of a *long arrest of culture* of the reasoning

faculties and of the will, and the cultivation of the sentimental traits and affections. I think the present era has
demonstrated that, with equal opportunity (in spite of this
long arrest of culture), women are carrying off the honors in
every department of intellectual labor.

There are those who may object to this exposition of the
superiority of woman, and may advance the argument that
man and woman are a whole—each the complement of the
other; that the female of any species is but the half of that
species. This would prove nothing more than an inexact
way of speaking. Those who care to trace the evolution of
species will find that the female takes rank as a perfected
organism all along the line of progressive growth, and many
species of insects find very little need of the male. Burns
was scientifically correct when he wrote, "His 'prentice hand
He tried on man, and then He made the lassies, O." Some
of those poetic chaps write wonderfully prophetic and scientific at times—all unconsciously, it would seem.

The question of *quality* always takes precedence in Nature.
Until the prevalent erroneous opinion in regard to woman
is changed, and her true rank in Nature accorded her, and
the mother reign a queen in the home in fact as well as in
name, we shall continue to have the miserable, distorted
specimens in the shape of children, who fill our homes with
sickness and suffering, and who drag out a wearisome existence—made such, in many instances, during their pre-natal
life, by the enforcement of hurtful laws and commands im-
posed upon the mothers by fathers ignorant of all physiological or hygienic law, or by an intrusion of masculine will and
passion at times when Nature cries out against such infraction of her laws, which are and should be taught to the
world as divine. That sugar-coated fallacy which is dropped
into the ears of unthinking women—that "she who rocks the
cradle rules the world"—will then be a living reality, and
the home will then be worth acknowledging as "woman's
empire"—will be something to admire and be proud of—
something that the sons of woman can rejoice in; and not,

as now, a place often, too often, made hideous by the wholesale use of intoxicating liquors by the fathers, "the heads of
the family," which render the sanctuary of home often unsafe, and a precarious shelter for mothers and children.

Thinking women are not gratified when they are told that
the present demoralized condition of society results from
their "ruling." It is no compliment, but a disgrace, if it
were true. Woman has no voice or influence in legalizing
the sale and manufacture of liquors—in licensing gambling
hells and houses of prostitution, as is done under some of
the European governments and in some parts of our own
country. Woman rocks the cradle, it is true, and there it
would often seem that her "ruling" and influence ended. If
a mother expend twenty-one years of her life in educating
her son, surrounding him with all that beautiful home-influ-
ence which she is led to believe is so potent in protecting
him, she will (in many instances) live to see all her years of
labor rendered abortive by the influence of the above-men-
tioned legalized abominations, which she is not allowed to
assist in abolishing, because she is taught that her influence
is "mightier in the home," but which she finds to be the case
only so long as she can bring it to bear personally upon her
children while in the home, and that outside influences,
which she is not permitted to assist in reforming, *rule* her
home, her children, and herself. When her true place under
the law and in society has been allowed her, her influence
will be mightier than the sword, and will assist man in protecting him even from himself and his vices. I feel that I
have performed a religious duty in thus setting before my
readers woman's true place in Nature according to natural
law.

We are fast approaching the age when mankind will cease
to live so much in instinct, and begin to live more in reason.
The progress of man mentally will be in exact accordance
with his progress physically; first, the stomach age, when the
inhabitants of earth dwelt in the plains and marshes, when
men were great only as eaters and getters of food, which at

that time Nature bestowed without effort on the part of man,
Next, the age of breathing, when man came to live in the
high places and mountains. Then came the evolution of the
muscle age, when man was a "mighty hunter." This exercise in the open air facilitated the development of the osseous
or bony structure; then men became tillers of the soil, cattle
herders, and shepherds. Here commenced the growth of the
brain age, when man had freedom from the nomadic life
which was entailed by hunting and fierce conflicts with wild
beasts. Opportunity came for reflection in the peaceful retreats of pastoral and agricultural life. Thought was evolved,
and thought gave birth to mechanism, and mechanism led
the way to scientific knowledge.

The laws of mechanics type the laws which govern the
universe—not only our little planet, but all the worlds—for
there is a correspondence of law which applied science, in
the form of mechanical instruments, has discovered and verified. The spectroscope, telescope, and other instrumentalities, have each given its quota of positive knowledge of the
constituents of those mighty bodies, and thus mind brooding
over matter is slowly, yet surely, bringing us to higher realms
of thought, reason, and justice.

When men choose their wives in reason and justice, the
race will take one grand step forward. If as much attention
had been paid to the rearing of children as has been given
the culture of plants, horses, or hogs, we should be immeasurably advanced already. On this topic, Quatrefages, in his
work on "The Human Species," remarks:

"Now, plants and animals have been studied for a much
longer period than man, and from an exclusively scientific
point of view, without any trace of the prejudice and party
feeling which interferes with the study of man. Without
having penetrated very deeply into all the secrets of vegetable and animal life, science has acquired a certain number of
fixed and indisputable results which constitute a foundation
of positive knowledge and a safe starting-point. It is there
that the anthropologist must seek the known quantities of

which he may stand in need. Wherever there is any doubt of the nature or significance of a phenomenon observed in man, the corresponding phenomena must be examined in animals, and even in plants. They must be compared with what takes place in ourselves, and the results accepted as they are exhibited. What is recognized as being true for other organized beings cannot but be true for man. This method is incontestably scientific. It is similar to that of modern physiologists, who, since they are unable to experiment upon man, experiment upon animals, and form their conclusions upon the former from the latter. In anthropology, every solution to be sound—that is to say, true—should refer man to everything which is not exclusively human, to the generally recognized laws for other organized and living beings. Every solution which makes, or tends to make, man an exception, by representing him as free from those laws which govern other organized and living beings, is unsound and false."

In continuing the subject of heredity from a scientific standpoint, I shall discuss it upon these principles and laws mentioned by Quatrefages, for my observations of a life-time confirm them.

Had the Rev. W. H. Murray written as wisely on child-rearing as he has on horse-breeding, he would have laid the world under lasting obligations. As it is, his book, entitled "The Perfect Horse," is one step toward the cultivation of the Perfect Man, as nearly all the rules he lays down for the propagation of fine types of horses would apply exactly to the same purpose in the human family. Let us examine some of Mr. Murray's rules for producing a perfect horse, and see if, when applied, they would not also result in creating more nearly perfect men and women. He observes:

"Make yourselves familiar with the history of the noted horses of your own country, and also of other lands. Make yourselves acquainted with their shape, size, peculiarity of going, character of temperament, and the ancestry whence they sprung. Study pedigrees, that you may know, by the union of what bloods, and the intermarriage of what families, great results have been attained. The reader will see that I demand no more than is universally admitted to be the condition of success in other branches of business. I do but demand this, and I lay it down as a law which executes its own penalty when transgressed, that he who breeds a horse, while ignorant of the correct principles of breeding, will breed a failure. If he ever make a success, it will be based on no broader or surer foundation than mere luck. Like produces like, and a fine-blooded colt must have fine-blooded parentage, and no one can escape its application. Luck has nothing to do with breeding; knowledge and a wise use of means can alone secure you what you desire. Never breed from an ugly tempered mare, for her colts will surely be like her, only worse. Depravity gets an earlier development in the child than it had in the parent."

This principle holds good in regard to talents and virtues as well as to vices. Inherited traits of every sort show earlier in life where they are inherited from parents who possessed such naturally. The biographies of all great actors, *litterateurs*, jurists, and orators prove this. Mr. Murray says further:

"Lastly under this head, see to it that the mare selected for the stud be in perfect health; feel that there is no exception to this, for every trace of disease in the blood of the dam will from necessity be imparted to the foal—the embryo will from the very beginning be tainted with disease, so true it is that unhealthiness is often bred out of the dam and into the foal. The colt is worthless, but the mare is cured. The disease left the mother and entered into the offspring, as is the case often in the human species."

There is not in the foregoing a single principle mentioned which, if applied to the begetting of the human race, but would raise it immeasurably in the scale of nobility. I leave to the common sense of the reader if this is not correct.

One prolific cause of the deterioration of the race is found in the marriages of first cousins. A large percentage of the progeny of such marriages are either deaf and dumb, idiotic, scrofulous, or epileptic. Tabulated statements, which the reader can find in various medical works, give the percentage. The cause of this deterioration in these consanguineous relations is that their constitutions are too nearly alike; in other words, the parents resemble each other too nearly in form and in the general constituents of organism. Results point to this as a cause, and such unions should be avoided. The danger is not so great where the pair are very unlike, or who resemble ancestors of opposite sides of the family who were very unlike, particularly if quite free from constitutional disease. This law holds good also in cases where persons, not relatives, intermarry who possess very nearly the same forms and systems of functions, in nearly the same proportions. They are alike by nature, yet not by blood ties.

Given, two parents possessed of the same proportions of the Osseous and Brain systems in excess. The result will be nervous, bright, weakly children, unless some fortuitous and hidden law, of atavism for example, intervene, and the offspring receive the balancing effect of a modicum of some other system from some remote ancestral type. This seldom occurs, yet often enough to admonish us to study pedigrees especially where marriage is intended, and also to obtain the assistance of a first-class physiognomist. Being too much alike has the effect of weakening the offspring, inasmuch as it intensifies all constitutional defects, and gives too large a preponderance of certain systems and organs, and thus produces an unbalanced condition. And yet where the parents are exactly opposite in form, color, and systems of functions, this also breeds inharmony and disaster. There never was a more fallacious saying than that which passes current—viz., "that opposites should marry"—and those who make this observation can give no reason in its support. Such marriages create discord between the parents. What, then, can we expect will be the condition of their progeny? Even in cases of parents well mated, temporary discords between them have such an effect upon the mind of the mother during the pre-natal existence of her offspring as to result in inharmonious, unhappy, weak, and deformed children.

Lord Byron, the celebrated poet, was born deformed—the result, it is thought, of violent fits of anger on the part of his mother before his birth. She had inherited this proclivity to violence from ancestors who were at times half insane with anger; thus she received this vice in an intensified form, and the club-feet of Byron told the story of his mother's and his ancestors' habits.

If temporary discords produce disastrous results, what might not be the result to children whose parents 'lived in one ceaseless conflict, owing to the utter inability of the incongruous elements of their organisms to harmonize? Such natures cannot mingle any more than oil and water; in such a struggle, the weaker one is the chief sufferer. It is like rubbing a piece of silk against a rock; the silk is destroyed, while the rock will remain unharmed. Such partners may neither be vicious, but will become so if they remain long in such unequal yokedom. Each might find his natural and harmonious mate if the laws of Physiognomy were understood and regarded, and thus untold misery would be spared to children and children's children.

As society is at present constituted, the greater share of the responsibility of the selection of types from which to reproduce devolves upon man; for, according to custom, man has the privilege of choosing; a woman must marry the man who asks her, or not at all; and, being in the majority of cases financially dependent, generally accepts the man who can provide for her, without regard to the fitness of the union in regard to age, form, health, or any other consideration but the mere fact of ability to provide.

Woman is far more intuitional than man. This has always been conceded, and Physiognomy proves its truth. If this wondrous intuition could be allowed free scope in the matter of the choice of husbands, an almost entirely different allotment in marriage would be the result; and this one step would add immeasurably to the improvement of the race, for intuition, in its highest development, gives the most profound insight into human nature; hence, woman is pre-eminently endowed with the power for choosing the best types of man for perpetuation. But as long as this most glorious gift of woman's nature is subordinated to a *struggle for existence*, this one important and powerful factor in the culture of the race will be almost entirely closed against it. If woman were as free to choose her husband as man is now free to choose his wife, I believe all concerned would be the better for it. In that "good time coming," which we all hope for, possibly this with other methods of regeneration will supplant those now in vogue.

History shows that the highest types of the race are found where the monogamic marriage prevails, and where the home is inviolable. The welfare of the family depends greatly on the continuous efforts of the same parents. Human nature, in order that it should be conservative, is somewhat selfish, and, naturally, a parent will provide and do for his own offspring with more wisdom and affection than for the children of another. It is therefore highly important, not only for the welfare of the children, but for harmony between the parents, that there should be an adaptation in the parents for each other; and this should be ascertained *before marriage*. A skillful physiognomist could determine this question, as well as their fitness to become parents, and, also, if the union would be productive of superior offspring. It is the solemn duty of every one intending to enter the high and holy estate of matrimony to discover, by scientific knowledge, if the conditions attending this desire are righteous; that is to say, in accordance with morality, physiological law, and the laws of heredity and transmitted quality and types. I advise my readers to procure every work on heredity that can be procured, especially one entitled "The Human Species," by the eminent A. de Quatrefages; also, one by Francis Galton, entitled "Hereditary Genius." They will be useful to all, especially those intending to become parents.

Quatrefages, whom I shall have occasion to quote often in

this chapter, states that "every race is a resultant, whose components are partly the species itself, partly the sum of the modifying agents which have produced the deviation from the type." Again he remarks: "Like all animal and vegetable species, the human species can vary within certain limits; like plants and animals, man has his varieties and races, which have appeared and been formed by the action of the same causes. In the human kingdom, as in the two other kingdoms, the first causes of variation are conditions of life and heredity. In phenomena of this kind, conditions of life act as the supreme ruler; if they vary, they become modifying agents; if they remain constant, they become agents of statilization. Heredity, which is essentially a persevering agent, becomes an agent of variation when it transmits and accumulates the modifying actions of the conditions of life."

Elsewhere he says: "Man does not subject himself to the selection which he applies with so much success to animals and plants, in his species; therefore, the extreme variations which are obtained elsewhere are not produced. It is thus easily explained why the limits of variation are not so extreme with man as with domesticated and cultivated races. But if, for some motive or other, he were to apply the process of selection to himself, we should not have to wait long for the result. By marrying the tallest women to the giants of their guard, Frederick William and Frederick II. had created, at Potsdam, a real race distinguished for its tall stature. In Alsace, a duke—de Deux Ponts—who imitated the Prussian sovereigns obtained the same results."

This exposition of heredity by Quatrefages corroborates the experience of all observers in this direction; and all who love their kind would desire, I am certain, that a system of selection in the human family might be put in operation which would be in the interests of morality, and, at the same time, tend directly to the ennobling of the species. By studying the pedigrees of families, and by making surrounding conditions favorable to the advancement of offspring, it is proved incontestably that not only can present varieties be improved, but also that new races can be evolved by intelligence in crossing existing types.

The conditions of life required for the better development of the race are, first, hygienic; second, obedience to moral law; in other words, such treatment of the several systems of the body as is in accord with physiological law. I have shown, in this system of Physiognomy, that morality proceeds directly from a proper development (at birth) of the several systems and their accompanying organs, and also by strict adherence to the laws regulating their operation. Morality is not a sentiment—not a mental conception, merely, of right and wrong—but is part and parcel, as I have heretofore shown, of the physical system. It is true that the mind must take cognizance of the operations and laws of the various organs of the body, and the will must be exerted to compel obedience to understood law; a knowledge, alone, of moral law makes no man moral; it is his adherence to those laws. If a man is born unbalanced in his nature, he can, by attention to hygienic law, neutralize in a great degree this abnormal condition.

If the Vegetative system predominates over all the others, causing the individual to be slow, sluggish, dull of thought, given to ease and sensuality, all these disadvantages may be overcome in a degree by a persistent effort on the part of the individual, or with the assistance of parents or interested friends. Great changes may be effected in a few years by denying one's self somewhat in the matter of food, sleep, and the immoderate use of liquids. It is the same with all the other systems; they can all be added to or reduced. In the chapter on "Hygiene," directions how to make these changes were given.

In endeavoring to trace the laws of Nature to their origin, man collects a great variety of knowledge without, perhaps, attaining the desired object; yet it is only in thus doing that he can gather sufficient knowledge of laws to advance the interests of humanity. Through the efforts of stock-breeders to produce better types of animals, and by experiments in the animal kingdom performed by scientists and naturalists, we have accumulated a vast store of knowledge in regard to the operation of the laws of heredity. It has been proved that in the human family many results have been observed similar to those occurring in the animal kingdom, where the selection was instinctive, and not designed by reason and intention. The same qualities and traits, as far as they have been observed, present similar results in both the animal and human races.

There are objections raised by the unthinking against the application of all the laws of propagation of the animal species to the procreation of man. I refer to the fact that many, having observed the sad result to offspring of the intermarriage of near relatives, conclude that the method of crossing of near relatives, as is practiced without detriment, and often with the best results, in the animal species, cannot be applied to the human family. They infer that, as this works disastrously, none of the other laws which are found to succeed with animals can be safely trusted to produce better types in the human family. With very few exceptions, the most able scientists are unanimous in the belief that the reason why the human family cannot safely apply this method is owing to the fact that none of the human family are exempt from disease or predisposition to disease derived from some of their ancestors. Now, the union of two persons predisposed to the same disease or class of diseases—those which all the members of the same family would probably exhibit—would, in their offspring, develop and intensify in its highest activity such diseased conditions and predispositions. The animal races, living more nearly in accord with the laws of Nature, present very rare instances of predisposed disease, and where this is observed the stock-breeder prevents reproduction from such source. Animals do not have the almost universally diseased conditions which prevail in the human species; they would be of no use to man for food were this the case, and could not reproduce the wonderfully beautiful and healthy specimens which we find among them

Among the laws of heredity is one named "atavism," or, as some express it, "reversion," or taking back. This law operates to transmit to distant descendants the diseases, vices, virtues, and talents of some remote ancestor, without exhibiting its peculiarity in the intervening generations; that is to say, a trait or disease will often overleap two, three, or more generations, and then make its appearance. This shows the importance of knowing the pedigree for several generations, at least, where the scientific improvement of the race is desired. We have all sorts of proof of the activity of this law. In cases where there have been infusions of African blood several generations back, it has produced the predominant traits, and some degree of color and features, in the most unexpected manner, after several generations of absence. This same law operates in the transmission of diseases. Cancer is often inherited in this manner; insanity also descends to children's children, and beyond these even. Mr. Sedgwick, a noted writer on stock-breeding, says:

"In the well known case of George III. the insanity was transmitted in the male line by atavic descent, from a male ancestor eight generations back, in whom not only the insanity, but many other of the well known characteristics were exactly repeated."

The most eminent talents, as well as the most moral characteristics, are transmitted directly and through ancestral influence. The science of heredity is yet in its infancy. If the time ever arrives when men and women shall be sufficiently enlightened and sufficiently unselfish, the attempt to regenerate the race by scientific methods will be made, and the laws and rules which have produced such satisfactory results in improving the lower animals will assist in raising man to a higher scale of development, added to the laws yet to be discovered, which relate exclusively to man and his more highly specialized organism.

Physiognomy, as taught by this system, is a powerful ad-

junct to race improvement. It not only teaches the meanings of the several systems and forms of the body, but also
the several qualities, traits, virtues, vices, weaknesses, talents, genius, and predispositions to health, disease, and longevity. The chapter on "Hygiene" gives the rule for the
improvement of each particular system by diet and hygienic
living. It is not within the scope of this work to give an
exhaustive *resume* of all the topics treated of, but its intention is to call the reader's attention to these subjects, leaving
somewhat for individual research and observation. The
most notable influences on heredity, and which play the most
important part in the propagation of the race, are those
which have been noticed in the preceding pages. These include inherited talents, diseases, vices, forms, colors, and
transmitted quality, as well as climate, food, water, clothing,
and social advantages. These are briefly alluded to, yet in a
manner in which I trust will call the attention of my readers
to their importance to themselves and their progeny.

CONCLUSION.

Looking far down the vista of the ages, we find coming
up from simple plasmoid substance, through gradational
forms, the wondrous organism of man, with all his complex
and varied powers. The earth, too, we find has kept abreast
in its evolution with the ever-changing wants and conditions
of organic life. It is logical to infer, reasoning from every
lesson of the past, that changes will continue, that no part
of Nature's domain will either retrograde or come to a stand-
still. We have also every proof, reasoning from experience,
that progress in man's nature will continue, and that the conditions requisite for that progressed nature will be evolved
in accordance with the ratio of his progression. I believe
that this world will witness greater and more striking changes
than have already occurred. Scientific knowledge having
attained such impetus as now exists will assist in carrying
forward this evolutionary movement of man with rapid strides.

Can any one imagine even the possibilities of ultimate growth? Is there any law, which has been observed in any department of Nature, that has ever exhibited a limit to action?—that is to say, does any element or particle ever cease being an agent or actor? I think no one can define the ultimate limit and action of natural forces.

What, then, must be our conclusion in regard to man's progress? It is simply illimitable. The human mind can no more comprehend the extent of that upward growth than it can the idea of eternity. My belief, born of and strengthened by the observation of the laws of God as exhibited in the laws of Nature, brings me to the conclusion that the three great ruling laws in Nature, the Chemical, the Architectural, and the Mathematical, will continue the grand plan of evolution, or progressive development, through the ever varying conditions of life-growth, and that as long as man exists as a conscious entity his face will register their action in his organism.

CONTENTS.

tecture, and Mathematics—find their representation in the anatomy of
Man, and the signs indicating these powers in his face....The human
face, like the dial on the clock, the exponent of all within the body....
The high importance of correct physiognomical knowledge ...Examina-
tion of sub-basic principles—viz., Form, Color, Texture, Size, Quality.
Shape of the Nose ...Harmony....Compensation....Illustrations of the
law of compensation....The five systems of functions which create the
various forms of Man, as well as his mental characteristics—the Vegetative, Thoracic, Muscular, Osseous, and Brain and Nerve systems....Fac-
ulties found in the Chemical or Vegetative division—Conscientiousness,
Firmness, Alimentiveness, Benevolence, Amativeness, Love of Children,
Mirthfulness, Approbativeness, Modesty, Self-esteem....All these faculties sustained mainly by chemical action....Architectural division of the
body and face, shown by predominance of the Thoracic, Muscular, and
Osseous systems, includes all the principles of mechanical forces—viz.,
lever, pulley, valve, etc....Faculties located in this division are Cautiousness, Secretiveness, Force, Resistance,·Hope, Analysis, Imitation,
Ideality, Sublimity, Human Nature, Constructiveness, Acquisitiveness,
Veneration, Self-will, Credenciveness, Observation, Memory of Events,
Form, Size, Weight, Color, Order, Calculation, Locality, Music, Language....Mathematical division shows action of Brain and Nerve system
and includes faculties named Time, Causality, Comparison, Intuition....
All the systems interrelated....The functions of the body produce mental effects....Moses's comprehension of facial indications shown in the
Bible....Index of body and mind in the face....False systems of metaphysics responsible for low grade of humanity....System of checks and
balances in body and mind, illustrated by Edgar A. Poe, Horace Greeley....Utility of Physiognomy—important to mothers, friends, partners
for life, and in choosing employments....Keen analysis required in reading character....Amativeness gives constructive talent—proof in faces
of Dickens, Rubens, etc....Laws of Heredity taught by Physiognomy.
Man's scant knowledge of himself....No progress in Government until
scientific knowledge of Man is universal.

All form throughout Nature symbolic of character....Advanced Phrenolo-
    gists accept this law....Practical value of Physiognomy....Its superior-
    ity to Phrenology....The five systems of functions create every type of
    character....The Vegetative, the first to evolve, shows the functions of
    reproduction, assimilation, imbibition, excretion, secretion, respiration,
    absorption, circulation—being the functions common to vegetable and
    animal existence....How to distinguish the Vegetative system....The
    intestines, liver, spleen, bladder, kidneys, pancreas, and organs of repro-
    duction included in this system....Moral sense needed to conserve the
    functions and faculties of this department....The Architectural division.
    Description of the Thoracic system....The progressive development of
    Man's organism follows the same course as that of the lower animals.
    The fish types the vertebrate classes....The lungs and heart necessarily
    correlated....No great men with large brains and small lungs....All the
    celebrated warriors, discoverers, orators, and commanders possess large
    lung and heart systems....Description of the Muscular system....Great
    size not evidential of great strength....Fair quality and large endow-
    ment of the muscles needed for artistic and mechanical pursuits... No
    mechanics found in the Vegetative system predominant....All the prin-
    ciples of natural mechanics represented in the muscles—the several lever
    powers, pulley, ball and socket, valves, mixed joints... The Bony system.
    Signs....Constituents and nature of bone....Its firm nature gives integ-
    rity—also scientific and mechanical ability....Choosing trades from the
    conformation of the body....Mathematical division....Description of
    the Brain system....An excess of Brain and Nerve system tends to rea-
    son and judgment—gives purity, refinement....Signs of Causality,
    Comparison, Time, and Intuition manifested in this department.

Chemical division....The voice, form, walk, gesture, all indicative of character....Primary use of bodily functions....The mind acquiring new faculties....Science taking the place of superstition....Physiognomy to assist in civilization .. Large features exhibit more power than small ones....Character spread all over the body....Locality of signs in the face—Conscientiousness, Firmness, Alimentiveness, Benevolence, Generosity, Sympathy, Amativeness, Love of Children and Animals, Mirthfulness, Approbativeness, Friendship, Self-esteem, Modesty....The signs found in the Architectural division....Phrenology indebted to Physiognomy for its facial signs....What Mr. Fowler says of the face.... Conscientiousness not a religious faculty, but a moral one....Force, Secretiveness, Resistance, Hope, Cautiousness, Analysis, Imitation, Ideality, Taste or Imagination, Sublimity, Human Nature, Constructiveness, Acquisitiveness, Veneration, Executiveness, Self-will, Credenciveness, Observation, Form. Size, Weight, Color, Physical Order, Mental Order, Calculation, Locality, Memory of Events, Music, Language.... Mathematical division....A good proportion of the Brain system needed for mathematical computation, surveying, navigation, astronomy....Is Man's allotted time on earth a matter of mathematical certainty?....Sir John Herschel's statement of the mathematical nature of all organisms. Signs for Time, Causality, Comparison, Intuition.

Form and Size.....Evidences in Nature of uniformity of methods of action. Resemblance in form between men and animals result in similar characteristics....The duty of parents to understand and teach Scientific Physiognomy....Size not always indicative of strength....Quality before quantity....True greatness not denoted by size of the nose alone.... Inherited quality determines power....Quality shown by fine clear skin and fine hair; the same law applies to animals....The law of compensation....Illustrations in the eyes of insects, wings of bats, elephant's proboscis....The action of this law in Man's physique; also, in his mental construction....Law of proportion....The artistic law does not agree with the scientific....What Lavater says of the Greek profile....Winkleman's opinion....Scientific interpretation of proportion....Law of harmony....Perfect harmony in all of Nature's works....Bodies suited to the minds accompanying them....The external of Man always harmonizes with the interior construction....Law of color....Certain amount of coloring pigment essential to health and normal action of the functions.

The present age of invention a revelation....Why does every force in Nature act a double part?....The uses of jealousy, revenge, suspicion, force, all necessary in the present condition of mankind .Rationale of love, jealousy, revenge, suspicion, secretiveness, self-conceit The human race propagated instinctively, the same as animals. Theory of the action of love—its assistance in constructive art.. Physiological and hygienic law should be observed as a religion ..Theory of the action of jealousy—found only with unbalanced and undeveloped natures....Revenge found mainly among the darker races—never among well balanced people....Secretiveness, used to conceal something. ..Animals use secretiveness in order to avoid and prey upon each other....Suspicion arises from defective reason, practicality, or friendliness....Anger unjustifiable and useless....Righteous indignation the only legitimate action of this trait....Selfishness, self-protective and useful to a degree too much selfishness indicates an undeveloped nature....Excess of any

dominant....Intermarriage of cousins—to be avoided in some cases...
Law of ancestral inheritance....Opposites should not marry....The
most responsibility in reproduction attached to Man—why....Woman's
intuition a powerful adjunct in scientific culture of the race....The monogamic marriage the best—the reason why....What Francis Galton
says....The conditions required for the better development of mankind.
Morality not a sentiment....The reason why near relatives in the human
race should not reproduce, and why near relatives in the animal kingdom may....How George III. inherited insanity....Talents and vices
transmitted for generations—physical defects, also....Physiognomy a
great guide to the elevation of the race by scientific methods. ..The
most important influences in heredity ..Conclusion.